ELECTROMAGNETIC MEASUREMENTS IN THE NEAR FIELD

ELECTROMAGNETIC MEASUREMENTS IN THE NEAR FIELD

Pawel Bienkowski
Hubert Trzaska

Second Edition

SciTECH
PUBLISHING, INC.

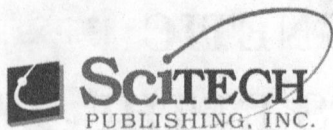

SciTech
PUBLISHING, INC.

Published by SciTech Publishing, Inc.
911 Paverstone Drive, Suite B
Raleigh, NC 27615
(919) 847-2434, fax (919) 847-2568
scitechpublishing.com

Editor: Dudley R. Kay
Production Manager: Robert Lawless
Typesetting: J. K. Eckert & Company, Inc.
Cover Design: Brent Beckley

This book is available at special quantity discounts to use as premiums and sales promotions, or for use in corporate training programs. For more information and quotes, please contact the publisher.

10 9 8 7 6 5 4 3 2 1

ISBN: 9781891121067

Library of Congress Cataloging-in-Publication Data

Cataloging in Publication Data Applied For

Contents

Preface

The importance of electromagnetic field measurement is greater now than ever before. This situation is only going to increase because of the increase in devices emitting radiation and the subsequent need for improved knowledge of sources in EMC. Near-field measurement is also very important for biomedical applications where the knowledge of fields is important for the protection of personnel. Far-field measurements are generally well understood and well documented. However, this is not the case with near-field measurements, and the sources and causes of errors in these measurements can lead to results that indicate only a passing resemblance to the real values.

Pawel Bienkowski and Hubert Trzaska have produced a book that will be invaluable to those designing or using near-field measurement probes. This book shows WHAT should measured, WHY it needs to be measured, and HOW it should be measured. It contains detailed designs and analysis of the circuits to be used.

Pawel Bienkowski and Hubert Trzaska are both internationally respected researchers and practitioners. This book is based on years of experience and builds on the previous edition. It blends absolute academic rigor with satisfying a practical need. It clearly has a place on the bookshelves of practicing EMC engineers and test and measurement engineers. It is also essential reading for those involved in environmental monitoring of EM fields and medical physicists who need to understand EM exposure.

Preface to the Second Edition

This is a revision of *Electromagnetic Field Measurements of the Near Field,* published in 2001 through Noble Publishing. Though we have streamlined the title, the proliferation of electronic devices and their resulting sources of electromagnetic fields (EMF) has dramatically increased. The misunderstandings and questions concerning EMF measurements, particularly in the near field, have led us to explain the phenomena and measurement techniques in more detail and with greater clarity. Also, a remarkable number of errors occur in near-field measurements. Presenting an understanding of their causes and essence should be quite valuable to those who undertake such measurements and the interpretation of results.

A greater understanding of near-field EMF measurements and causes of errors will be particularly beneficial to readers concerned with these three areas:

- The design and manufacture EMF probes
- Measurements taken for labor safety and environment protection purposes, as well as research in the area of bioelectromagnetics
- Interpretation of measurement results for legal decisions bearing on standards, regulations, and compliance

In our investigation of meters currently available on the market, the authors have found they do not always meet certain metrological requirements for accuracy. Perhaps the manufacturers are indifferent or unaware of the conditions in which their products will be applied. On the other hand, users of the devices may not be using these meters correctly or may be unaware of how to test them to understand their limitations. It is our hope that this book will assist manufacturers and users alike to better understand the many variables that can cause errors and to recognize them when they occur. A misunderstanding or misinterpretation of the EMF measurement's accuracy is especially worrisome in biomedical investigations. According to the authors' estimates, many investigations do not fulfill the basic methodological requirements of obtaining accurate measurements and their proper interpretation.

In the legal arena, protection against unwanted radiation exposure is a matter of complying with radiation limitations. Yet these limits are frequently exceeded, either because measurements are inaccurate or are misinterpreted as to what a measurement range may actually mean. Thus there is the very serious necessity here of understanding the measurement results. For instance, 10 V/m means that the real exposure is within range 8 to 12 V/m or wider.

As we release this revised edition into the hands of readers, we would like to express our hope that it may be helpful not only within the explicitly presented areas but also within the broader range of electromagnetic compatibility, where EMF measurements are of primary importance.

The authors would like to invite the readers to present their suggestions for additional information and improvements, as well as pose questions or concerns related to the issues presented. All of them will be appreciated very much.

Pawel Bienkowski

pawel.bienkowski@pwr.wroc.pl

Hubert Trzaska

hubert.trzaska@pwr.wroc.pl

Wroclaw, Poland
September, 2011

Acknowledgments

This revision has been a long and difficult journey as we attempted to put our thoughts into clear and logical English for the widest possible readership. This goal is not so easy to accomplish by two authors in Poland! We are therefore very grateful for the assistance from reviewers and our publisher. Dr. Randy Jost (Utah State University), Mr. Andrew Drozd (ANDRO Computational Solutions), and Mr. Larry Cohen (U.S. Naval Research Laboratory) provided early reviews of the content and expression. Mr. Jeff Eckert (J. K. Eckert & Company, Inc.) then provided an edited preliminary version of the book in actual book format, complete with finished figures, so that it could be reviewed further. Our publisher, Mr. Dudley Kay, then arranged for the fine-tuning of the book by bringing it into the SciTech Series on Electromagnetic Compatibility and enlisting the aid of the series editor, Dr. Alistair Duffy (De Montfort University, UK). We thank Brent Beckley of SciTech for his work on the cover so that it shows exactly the kind of EMF measurement activity that is the subject of the book, and still it is also very attractive. And so the book has been an extraordinary team effort, and we give sincere thanks to all. Any remaining errors or lack of clarity are our own responsibility.

Chapter 1

Introduction

This book is concerned with near-field measurements and in this chapter, we would like to set the context of why this is necessary from a "human" point-of-view and introduce the field magnitudes that are important. The degradation of the natural electromagnetic environment is the forgotten price that must be paid for our inconsiderate enthusiasm for "industrial revolution." As a result, we are reaching a situation in which spending for the protection of the environment must sometimes exceed the investment in the systems causing the degradation.

The development of contemporary civilization is associated with the consumption of more and more energy, in forms that are utilized by technology, science, medicine, and our households. One form of the energy playing a rapidly growing role in everyday life is the energy contained in radio frequency (RF) currents and fields. In some applications (telecommunication or radiolocation), the energy is the final product, while in others it is an intermediate form, designed to be transformed, for instance, into heat. In all cases, there can be intentional or unintentional irradiation of part of that RF energy, and, as a result, contamination of the whole environment and interference over a wide frequency range.

The natural electromagnetic environment can be understood through the electric field (E), the magnetic field (H), and the electromagnetic field (EMF). The natural environment has been overwhelmed by global wireless communication systems and power systems based on alternating current. In such systems, the intensities of artificially generated fields exceed natural ones in most cases—even at long distances from the radiation sources. Recently, we have observed an increase in wire (fiber-optic) data transmission and global satellite systems to meet global telecommunications needs. Simultaneously, however, there has been an explosion in the popularity of wireless communication and data

transmission systems (wireless phones, cellular phones, radiotelephones, remote control and radio-frequency identification devices
[RFIDs], WLANs, Bluetooth). We also see EMF-generating devices
(microwave ovens, dielectric and inductive heating, computers, and
monitors), especially in households. These systems are causing the
entire global population to exist in an electromagnetic environment for
which the adjective *natural* has not been applicable for 50 to 70 years.

Among the distinctive features of this degradation of the natural
EMF environment, as compared to other forms of environmental pollution, are:

- It is a unique realm (telecommunications) in which the pollution is
  caused intentionally.

- Its pollution is widespread, acting immediately and on a global
  scale.

- The exposure of the people working in the vicinity of high-power
  sources (telecommunications) is much lower than that of people
  located near medium-power sources (industry, science, medicine,
  household) and even low-power sources (mobile communication).

- It is the single area in which there is a theoretical possibility of elim-
  inating pollution all at once, completely, and without any residue.

The last item, however, would require the elimination of all devices
and systems based on electrical energy. It would mean a return to a pre-
industrial era. Such a solution would not be acceptable to modern society. Therefore, we urgently need to arrange for the compatible
coexistence of electromagnetic pollution and society. Mechanisms of
electromagnetic interference are well known, and appropriate protection methods have been devised. However, the most controversial and
concerning aspect of this "pollution" is the biological effects of EMF
exposure.

Research into the biological effects of currents and electric and magnetic fields goes back many hundreds of years. This includes the beneficial applications in medical diagnostics and therapy as well as the
hazards these fields create for humans. The pace of this research has
accelerated as the development of the technology and techniques for
EMF generation has intensified [1, 2].

Electromagnetic fields are not generally detectable by organoleptic
methods except for a narrow frequency band humans refer to as light
and other defined biomagnetic sensing in animals. Thus, EMF detection, and all work and investigation related to the field, requires the use
of tools for detection and measurement. Moreover, EMF is not directly

measurable, so it is necessary to convert it to an another quantity that we can measure (e.g., voltage, heat).

EMF measurement in the far field (Fraunhofer zone) is generally less accurate than comparable measurements of other physical quantities. Determining the hazards created by exposure to EMF requires field measurements in the neighborhood of primary and secondary field sources as well as fields disturbed by the presence of material media in the measurement area. Our attention must be focused on the near field (Fresnel region). The near-field conditions cause further degradation of the near-field EMF measurement accuracy compared to those in the far field. These difficulties are sources of debate about the measuring equipment among its users and bring frustration to its designers. Although not considered in this book, an "Achilles heel" here is the accuracy of EMF standards. A standardized device cannot be more accurate than the standard used for the procedure. At present, the accuracy of a "good" EMF standard does not much exceed 5 percent [5].

This book is devoted to the specific problems of EMF measurements in the near field and to the analysis of the main factors limiting measurement accuracy. It focuses on the measurements required for regulations and standards associated with labor safety and general public protection against unwanted exposure to EMF. While these issues represent the experience of the authors, almost identical metrological problems exist in electromagnetic compatibility (EMC). The analyses presented here make it possible to estimate the importance and the role of various factors involved in specific conditions of a measurement, as well as evaluation of commercially available meters (and their manufacturers).

Measurements related to surveying or/and monitoring require the use of quantities that are relatively simple to measure. They require meters that provide acceptable reliability and accuracy. However, measurements are sometimes performed in difficult field or industrial conditions. For practical reasons, we will limit ourselves to only the units whose abbreviations form the acronym HESTIA—the goddess of the fireside [6]. These quantities, derived quantities, and several constants useful in further considerations are set up in Table 1.1.

In free space and in nonmagnetic media, the magnetic flux density, B, is equal in value to the magnetic field intensity, H (multiplied by the permeability of free space). This has resulted in some meters being calibrated in B-units. This is particularly true of meters designed for the measurement of magnetostatic or quasi-static (resulting from VLF) fields. Table 1.2 provides a straightforward comparison of units.

Table 1.1 Quantities Representing EMF and Their Units

Quantity	Symbol	International unit (SI)
Magnetic field strength	H	amperes per meter [A/m]
Electric field strength	E	volts per meter [V/m]
Power density	S	watts per sq. meter [W/m^2]
Temperature	T	kelvins [K]
Current intensity	I	amperes [A]
Magnetic flux density	B	tesla [T] = 10^4 gauss [G]
Current density	J	amperes per sq. meter [A/m^2]
Specific absorption	SA	joules per kilogram [J/kg]
Specific absorption Rate	SAR	watts per sq. meter [W/kg]
Conductivity	σ	siemens per meter [S/m]
Permittivity	ε	farads per meter [F/m]
Permittivity of vacuum	ε_o	$\varepsilon_o = 8.854 \ 10^{-12}$ F/m
Permeability	μ	henrys per meter [H/m]
Permeability of vacuum	μ_o	$\mu_o = 12.566 \ 10^{-7}$ [H/m]

Table 1.2 Mutual Correspondence of H-field Units in a Nonmagnetic Medium

A/m	796	80	8	0.8	80 mA/m
Gauss [G]	10	1	0.1	10 mG	1 mG
Tesla [T]	1 mT	0.1 mT	10 µT	1 µT	0.1 µT

The parameters of field strength meters are defined in several national and international standards [8–11]. The standards are only partly related to the specific types of required near-field measurements. From the standpoint of this book, the standards also define more parameters than we require. On the other hand, the most essential parameters are presented without any comments that would make it possible to analyze the measurement conditions and the domain in which the meter may be successfully applied. The latter is easy to understand, as it is not the subject of these documents. In order to better introduce metrological needs, Table 1.3 lists the current EMF exposure limits in accordance with standards in Poland [11]. In Tables 1.4 and 1.5, we present some fraction of the European regulations worked out by a competent international team [11] and then accepted as recom-

Table 1.3 Exposure Limits under Polish Environmental Regulations [11]

Frequency range	E [V/m]	H [A/m]	S [W/m^2]
Static field	10,000	2,500	–
0–0.5 Hz	–	2,500	–
0.5–50 Hz	10,000	60	–
0.05–1 kHz	–	3/f	–
0.001–3 MHz	20	3	–
>3–300 MHz	7	–	–
>0.3–300 GHz	7	–	0.1

Table 1.4 Permissible Exposure Levels in Accordance with the IRPA Proposals [12]

Frequency range	E [V/m]	H [A/m]	B [T]	S [W/m^2]
>0–1 Hz	10 000	3.2×10^4	4×10^4	
>1–8 Hz	10 000	$3.2 \times 10^4 /f^2$	$4 \times 10^4 /f^2$	
>8–25 Hz	10 000	4 000/f	5 000/f	
>0.025–2.874 kHz	250/f	4/f	5/f	
>2.874–5.5 kHz	87	4/f	5/f	
>5.5–100 kHz	87	0.73	0.91	
>0.1–1 MHz	87	$0.23/f^{-1/2}$		
>1–10 MHz	$87/f^{-1/2}$	$0.23/f^{-1/2}$		
>10–400 MHz	27.5	0.073		2
>400–2000 MHz	$1.375\,f^{1/2}$	$0.0037\,f^{1/2}$		f/200
>2–300 GHz	61	0.16		10

mendations for members of European Community [13]. Only uncontrolled environment limits are presented below to illustrate the metrological needs. In Tables 1.4 and 1.5, frequency (f) is in the units indicated under the column heading, "Frequency range."

In the United States, two proposals were recently made [15, 16]. Both of them, with regard to permissible exposures, are similar to the U.S. national standard ANSI/IEEE C95.1–1992 (Table 1.6). Somewhat different levels of exposure are given in proposals presented by the American Conference of Governmental Industrial Hygienists (ACGIH) and related to the controlled environment [15].

Table 1.5 Permissible Current Intensity in a Hand or a Foot [12]

Kind of risk	Frequency range	Conduction current [mA]
Professional	1 Hz–2.5 kHz	1.0
	2.5 kHz–100 kHz	0.4 f
	100 kHz–100 MHz	40
General public	1 Hz–2.5 kHz	0.5
	2.5 kHz–100 kHz	0.2 f
	100 kHz–100 MHz	20

Table 1.6 Exposures Permitted by American Standards [15, 16]

Frequency range	E [V/m]	H [A/m]	S (PE, PH) [W/m^2]		T_{AV} (E,S)	(H)
3 kHz–100 kHz	614	163	10^3	10^7	6	6
100 kHz–1.34 MHz	614	16.3/f	10^3	$10^5/f^2$	6	6
1.34 MHz–3.0 MHz	823.8/f	16.3/f	$1800/f^2$	$10^5/f^2$	$f^2/0.3$	6
3.0 MHz–30 MHz	823.8/f	16.3/f	$1800/f^2$	$10^5/f^2$	30	6
30 MHz–100 MHz	27.5	$58.3/f^{1.668}$	2 9.4	$10^6/f^{3.336}$	30	$0.0636f^{1.337}$
100 MHz–300 MHz	27.5	0.0729	2		30	30
300 MHz–3 GHz			f/150		30	
3 GHz–15 GHz			f/150		90000/f	
15 GHz–300 GHz			100		$616000/f^{1.2}$	

Table 1.7 Permissible Currents Induced by EMF [15, 16]

Risk type	Frequency range [MHz]	Maximal current of both feet [mA]	Maximal current of a foot [mA]	Conduction current [mA]
Professional	0.003–0.1	2000f	1000f	1000f
	0.1–100	200	100	100
General public	0.003–0.1	900f	450f	450f
	0.1–100	90	45	45

In Tables 1.6 and 1.7, f = frequency in megahertz and T_{AV} = average time in minutes.

The above cited proposals of IRPA standards, as well as American ones, are based upon detailed studies of biomedical and physiological issues. Especially well founded are proposals of the IRPA [12], and analyses of the studies have been published [17–19]. Although *"the time between formulation of the proposals to their implementation may be as long as from the Acropolis construction to the proposals' formulation"*[19], nevertheless, just now, they may be useful for indicating the direction of further metrological needs.

The authors, as electronic engineers, have never reserved for themselves any right to suggest what should be the exposure limits. However, our personal opinion is that it is impossible to believe that the bioeffects are so precisely known that it was possible to propose standards with an accuracy to the third decimal point, not to mention the possibility of making field strength measurement with such accuracy! Aside from the controversy regarding the levels presented in the tables, the inclusion of these levels serves only as an introductory estimation of the EMF strength measurement range. More precisely, this establishes the upper limits of measured fields, since the lower ones could be below the noise level of the most sensitive meters. Perhaps exposure limits should be proposed by biologists and medical doctors or ecologists with physicists and engineers serving a more auxiliary role. The limits prepared in such a way may be a bit less precise, but they will probably be much more humanitarian. A trend in that direction has already been demonstrated [21, 22].

Having given an introduction to the domain of interest as well as the magnitudes of measurements required, the rest of the book addresses the elements of near-field measurements in turn:

- Chapter 2 defines the near field in relation to the far field.
- Chapter 3 introduces field measurement methods.
- Chapters 4–6 look at the measurement of the electric and magnetic fields as well as power density measurements in more detail.
- Chapter 7 addresses the issue of radiation pattern measurement.
- Chapter 8 describes important considerations limiting the accuracy of measurements.
- Chapter 9 introduces photonic approaches to field measurements.

References

[1] S. Szmigielski. The paths and the wilderness of the bioelectromagnetics (in Polish). *Proc. of the XIV-th Autumn School of PTBR*, Zakopane 1993, pp. 1–5.

[2] S. M. Michaelson, M. Grandolfo, A. Rindi. Historical development of the study of the effects of ELF fields. In: *Biological effects and dosimetry of static and ELF electromagnetic fields*, pp. 1–14. Plenum Press 1985.

[3] R. A. Waver. The electromagnetic environment and the circadian rhythms of human subjects. *Ibid.* pp. 477–524.

[4] A. S. Presman. *Electromagnetic field and the life* (in Russian). Moscow 1968.

[5] M. Kanda, D. Camell, J. P. M. de Vreede, J. Achkar, M. Alexander, M. Borsero, H. Yajima, N. S. Chung, H. Trzaska. International Comparison GT/RF 86–1 Electric Field Standards: 27 MHz to 10 GHz. *IEEE Trans.* Vol. EMC-42, No 2/2000, pp. 190–205.

[6] H. Trzaska. Power density as a standardized quantity. COST 244 WG Meeting Athens 1995, pp. 111–118.

[7] PN-77/T-06581. Labour protection against EMF within frequency range 0.1–300 MHz. EMF meters (Polish standard).

[8] PN-89/T-06580/02. Labour protection against EMF within range 1–100 kHz. EMF meters (Polish standard).

[9] Measuring equipment for electromagnetic quantities. Prepared by IEC TC 85 WG11.

[10] Radio transmitting equipment. Measurement of exposure to radio frequency electro-magnetic field–field strength in the frequency range 100 kHz to 1 GHz. IEC SC12C.

[11] The Decree of the Minister of Environment Protection of Oct. 30, 2003 related to the permissible EMF levels in environment and methods of the levels control (in Polish). Dz.U. No. 192/2003, pos. 1883.

[12] Guidelines on limits of exposure to time-varying electric and magnetic fields and to radio frequency electromagnetic fields (1 Hz–300 GHz). Draft, IRPA/INIRC 1994.

[13] Council recommendation of 12 July 1999 on the limitation of exposure of the general public to electromagnetic fields (O Hz to 300 GHz) (1999/519/EC).

[14] NATO Standardization Agreement 2345: Control and evaluation of personnel exposure to radio frequency fields.

[15] Threshold Limit Values for Physical Agents in the Work Environment. Adopted by ACGIH with Intended Changes for 1994–1995.

[16] Electromagnetic Fields (300 Hz to 300 GHz). *Environmental Health Criteria* 137. WHO, Geneva 1993.

[17] ICNIRP Guidelines. Guidelines on limits of exposure to static magnetic fields. *Health Physics,* No. 1/1994, pp. 100–106.

[18] ICNIRP Guidelines. Guidelines for limiting exposure to time-varying electric, magnetic and electromagnetic fields (up to 300 GHz). *Health Physics,* No 4/1998, pp. 494522.

[19] B. Kunsch. The new European Pre-Standard ENV 50166—Human exposure to electromagnetic field. COST 244 Working Group Meeting, Athens 1995, pp. 48–58.

[20] H. Trzaska. What about frequency independent standards? *Proc. 4th EBEA Congress,* p. 121–122, Zagreb, 1998.

[21] B. Eicher. Bioelectromagnetics: the gap between scientific knowledge and public perception. *Proc. 1999 Intl. EMC Symp.* Zurich, pp. 7176.

Chapter 2

The Near Field and the Far Field

The essential information for practical metrology is presented in this chapter, including a brief summary of the near-field properties as well as the basic equations and formulas related to fields generated by simple radiation sources.

2.1 AN EMF GENERATED BY A SYSTEM OF CURRENTS

Consider the calculation of the EMF at an arbitrary point, P, external to a volume containing arbitrary currents. Figure 2.1 shows a volume V, of arbitrary cross section, containing a system of arbitrarily oriented electric and magnetic currents, $\mathbf{J}$ and $*\mathbf{J}$, respectively. The volume V is surrounded by an infinitely large, homogeneous, isotropic, linear, lossless medium of permeability ε and permittivity μ. Electrical parameters are continuous on the boundary surface. The maximal linear size of the volume, V, is D (Fig. 2.1).

The $\mathbf{E}$ and $\mathbf{H}$ fields can be determined at an arbitrary point situated outside the volume, V, by solving Maxwell's equations for the angular frequency, ω [1]:

$$\mathbf{E} = \nabla \times \nabla \times \mathbf{\Pi} + j\omega\mu \, \nabla \times {}^{*}\mathbf{\Pi} \qquad (2.1)$$

$$\mathbf{H} = \nabla \times \nabla \times {}^{*}\mathbf{\Pi} - j\omega\varepsilon \, \nabla \times \mathbf{\Pi} \qquad (2.2)$$

where: $\mathbf{\Pi}$ and ${}^{*}\mathbf{\Pi}$ are electric and magnetic Hertzian vectors:

$$\mathbf{\Pi} = \frac{1}{j4\pi\omega\varepsilon} \int_{V} \mathbf{J} \, \frac{\exp(-jkr)}{r} \, dV \qquad (2.3)$$

11

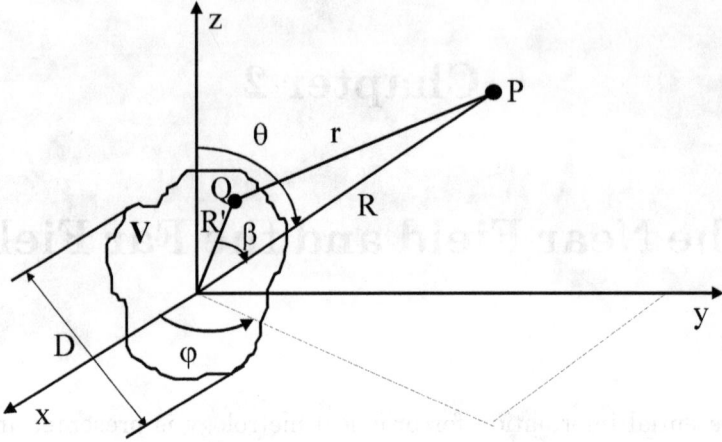

Figure 2.1   EMF in point P generated by currents in volume V.

$$^*\Pi = \frac{1}{j4\pi\omega\mu} \int_V {}^*\mathbf{J} \, \frac{\exp(-jkr)}{r} \, dV$$

$$(2.4)$$

where k = the propagation constant:

$$k = \omega \sqrt{\varepsilon\mu}$$

$$(2.5)$$

r = the distance from the observation point P to a general point
in the volume Q (R', θ', φ') so

$$\mathbf{r} = \mathbf{R} - \mathbf{R'}$$

with a resultant magnitude of:

$$r = \sqrt{R^2 + R'^2 - 2RR' \cos \beta}$$

$$(2.6)$$

where:

β = an angle between R and RR',

R = the distance from the observation point to the center of the
coordinate system,

R' = the distance from the general point to the center of the coordinate system.

Under the condition R' < R, the distance r may be presented with the use of a series expansion [Eq. (2.7)]:

$$r = R \left[1 - \frac{R'}{R} \cos\beta + \frac{R'^2}{2R^2} \left(1 - \cos^2\beta\right) + \frac{R'^3}{2R^3} \cos\beta \left(1 - \cos^2\beta\right) - \cdots \right]$$

(2.7)

If R >> D (where D is the maximal size of an arbitrary cross section of the volume V), it is possible to assume that r is parallel to R, so $r \approx R - R' \cos\beta$. Then:

$$\Pi_\infty = \frac{\exp(-jkR)}{j4\pi\omega\varepsilon R} \int_V \mathbf{J} \exp(-jkR' \cos\beta) \, dV$$

(2.8)

$$^*\Pi_\infty = \frac{\exp(-jkR)}{j4\pi\omega\mu R} \int_V {}^*\mathbf{J} \exp(-jkR' \cos\beta) \, dV$$

(2.9)

The index ∞ in the formulas indicates that they are valid for R >> D. In this case the spatial components of E and H are given by:

$$E_{\theta\infty} = -jk \left(Z\Pi_{\theta\infty} + {}^*\Pi_{\varphi\infty}\right)$$

(2.10)

$$E_{\varphi\infty} = -jk \left(Z\Pi_{\varphi\infty} - {}^*\Pi_{\theta\infty}\right)$$

(2.11)

$$E_{R\infty} = H_{R\infty} = 0$$

(2.12)

$$H_{\varphi\infty} = \frac{E_{\theta\infty}}{Z}$$

(2.13)

$$H_{\theta\infty} = \frac{-E_{\varphi\infty}}{Z}$$

(2.14)

where:

$\Pi_{\theta\infty}$, $\Pi_{\varphi\infty}$, ${}^*\Pi_{\theta\infty}$, and ${}^*\Pi_{\varphi\infty}$ = the spatial components of vector $\mathbf{\Pi}_\infty$ and ${}^*\mathbf{\Pi}_\infty$

Z = wave impedance of the medium:

$$Z = \sqrt{\frac{\mu}{\varepsilon}} = Z_0 \sqrt{\frac{\mu_r}{\varepsilon_r}}$$

$$\text{(2.15)}$$

$Z_0 =$ intrinsic impedance of free space:

$$Z_0 = \sqrt{\frac{\mu_0}{\varepsilon_0}}$$

Equations (2.10–2.14) allow us to find the far-field EMF components of an arbitrary system of currents in volume V. The field may be characterized as follows:

- The EMF in the far-field is a transverse field [Eq. (2.12)].
- At an arbitrary point the EMF has a shape of the TEM wave [Eqs. (2.13) and (2.14)].
- Vectors **E** and **H** can have two spatial components that are shifted in phase; as a result, the field is elliptically polarized.
- The dependence of the **E** and **H** fields on φ and θ is described by the normalized directional pattern that is independent of R.
- The **E** and **H** components are mutually perpendicular and related by the wave impedance of a medium.
- The Poynting vector $\mathbf{S} = \mathbf{E} \times \mathbf{H}$ is real and oriented radially.

To characterize the EMF properties in the far field, we have presented a straightforward solution of Maxwell's equations. To get a fully generalized solution of the equations, it would be necessary to take into account a number of additional factors including: the diffraction of a wave caused by irregularities in a nonhomogeneous medium, the dispersion and nonlinear properties of the medium, the anisotropy of the material, and the superposition of waves when nonmonochromatic fields are being considered. Such a solution has not been fully formulated. However, a fully general solution is not crucial for metrology and the following sections are based on the previous analysis.

2.2 THE FAR FIELD AND THE NEAR FIELD

The considerations presented above lead us to the description of several features that characterize the far-field. There are no actual discontinuities between the far field, the intermediate field, and the near field.

However, in order to adequately describe these regions, one of the criteria for their delimitation is presented below [2].

If we calculate the difference between the distance r given by Eq. (2.6) and its approximate magnitude given by the first two terms of the series in Eq. (2.7), we get an error distance. If we then multiply this by the wavenumber k, we have the phase error, $\Delta\Psi$. The limits to the use of the approximation R >> D is defined by the error and may be expressed in the form:

$$\Delta\Psi = k\left(\frac{R'^2}{2R}\sin^2\beta \; + \; \frac{R'^3}{2R^2}\cos\beta\;\sin^2\beta \; + \; \cdots\right)$$

(2.16)

At its maximum, 2R' = D, and this will result in the maximum value of the error given by following formula:

$$\Delta\Psi_{max} = \frac{kD^2}{8R} = \frac{2\pi}{N}$$

(2.17)

where: N = a number defining the acceptable inaccuracy of the phase front.

Usually, it is assumed that $N \geq 16$. So, substituting this in Eq. (2.17) and rearranging, we get:

$$R \geq \frac{2D^2}{\lambda}$$

(2.18)

The condition is widely accepted as the onset of the far field where D is now regarded as the maximum linear dimension of the antenna or other radiating structure. To illustrate this, consider two examples relating to antennas working at different frequencies:

- A parabolic reflector antenna of 3 m diameter working within the 10-GHz band.
- The tallest antenna long-wave transmitting antenna in the world, in Gabin (Poland), with a maximum height above 0.5λ, operating at 227 kHz (unfortunately, the antenna collapsed during guy-wire renovation, several years ago).

In both cases, the far-field can be obtained from Eq (2.18) and begins at distance above about 600 m away from the antenna. In the first case, R ~ $2 \times 3^2/0.03 = 600$ m, and in the second case R ~ $2 \times 661^2/1322 = 661$ m.

If, in our consideration, three terms of series [Eq. (2.7)] are taken into account with the other terms considered as vanishingly small, we have:

$$r = R - R' \cos \beta + \frac{R'^2}{2R} \sin^2 \beta$$

(2.19)

and then similar considerations to the far field are repeated, we obtain the following condition:

$$R \le \frac{D}{4} + \frac{D}{2} \left(\frac{D}{\lambda} \right)^{1/3}$$

(2.20)

Equation (2.20) gives the limit of the near field.

Figure 2.2 shows (after [3]) the field regions around an aperture antenna. Figure 2.3 illustrates the near- and the far-field boundaries as a function of r, D, and λ. The near field and the intermediate field are referred to as the Fresnel region (Fresnel zone), and the far field is referred to as the Fraunhofer region or the radiation field. When in close proximity to a radiation source, where the field may be assumed to be stationary, i.e., it is not radiating, the E and H fields are mutually independent. In this case, the use of the Biot-Savart law and Coulomb's law are usually assumed as sufficient. The field, in such close proximity to the antenna, is described as an induction field. Here, the imaginary part of the Poynting vector is dominant.

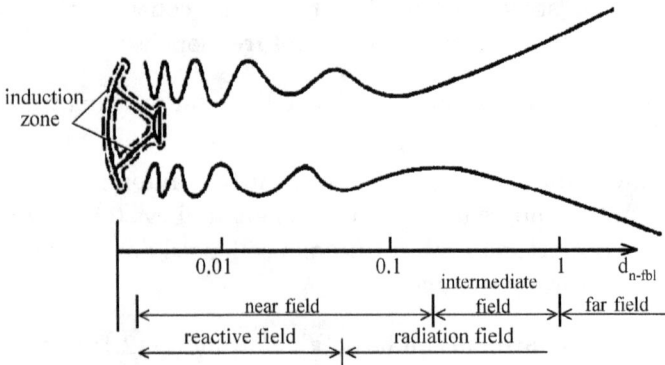

Figure 2.2   EMF in the proximity of an aperture antenna.

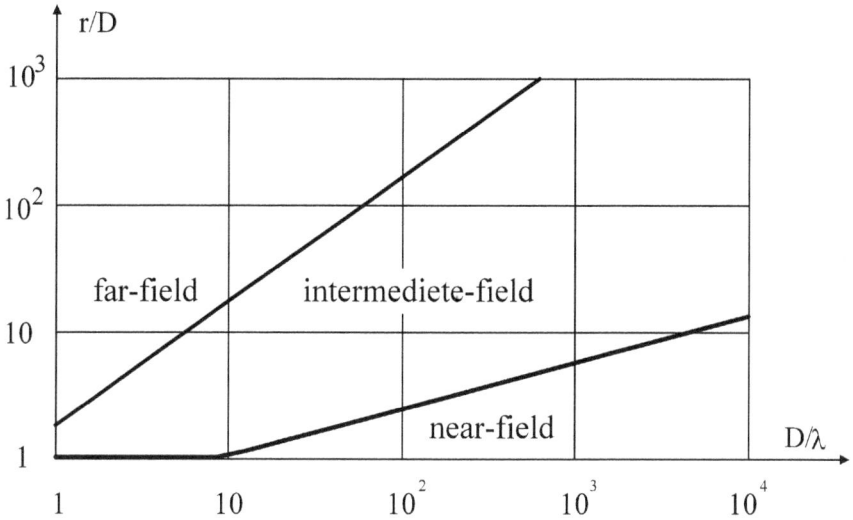

Figure 2.3   EMF near a source as function of r, D and λ; $d_{\text{f-fbl}}$—distance in relation to the far-field boundary limit as defined in Eq. (2.18).

The formulas introduced to define the near-field boundary [Eq. (2.20)] and the far-field boundary [Eq. (2.18)] require a word of comment. The series expansion given by Eq. 2.7 is true if R' < R, or more precisely, if:

$$R' < R \left(\cos \beta + \sqrt{\cos^2 \beta + 1} \right)$$

Although the conditions are not always fulfilled, Eqs. (2.18) and (2.20) are widely applied in the literature as definitions of the far-field and near-field limits. The accepted approximation here is a result of arbitrarily assumed permissible nonhomogeneity of the phase front N, rather than some inherent property of the fields themselves. At the boundary (or "border"), no actual discontinuity exists, and the expression *boundary* or *border* was introduced here so as to systematize the EMF parameters in the region surrounding a source. The near- and far-field definitions presented above are not the only definitions that exist. The boundary criterion may be based on, for instance, the convergence of the E/H ratio to Z_0, the Poynting vector to the electric (magnetic) power density, and others, but they are more difficult to systematize in the way that was done above, and the use of other criteria will give the boundary in the form of complex spatial function, while that given by Eq. (2.18) presents a sphere of radius R. Nevertheless, any criterion is based upon arbitrarily chosen values of a parameter, and the choice may be difficult to justify (e.g., why we accepted N = 16 instead of 15 or 17). As an example of another point of view, a definition of the EMF region boundary around radio station antennas is presented in Table 2.1 [4]. With no regard to the above, the role of the boundary in EMF metrology is limited only to warning that measurements performed at distances below R may be more troublesome and increasingly error-prone. Due to this, the authors have formulated a new definition of the near-field boundary: the near field is everywhere, where we perform EMF measurements. The definition is based upon longtime experience and observations of interfering and superposed fields, even at distances well in excess of R, due to multipath propagation and other phenomena, fields where a direct ray may not exist at all.

Table 2.1 Boundary of EMF Specific Regions around Radio Station Antennas [4]

Region	Region I	Region II	Region III
Region edges, measured from antenna where: λ = wavelength D = largest dimension of the antenna	$0 .. \max \begin{pmatrix} \lambda \\ D \\ \dfrac{D^2}{4\lambda} \end{pmatrix}$	$\max \begin{pmatrix} \lambda \\ D \\ \dfrac{D^2}{4\lambda} \end{pmatrix} .. \max \begin{pmatrix} 5\lambda \\ 5D \\ \dfrac{0,6D^2}{\lambda} \end{pmatrix}$	$\max \begin{pmatrix} 5\lambda \\ 5D \\ \dfrac{0,6D^2}{\lambda} \end{pmatrix} ..\infty$
$E \perp H$	No	Effectively yes	Yes
$Z = E/H$	$\neq Z_0$	$\approx Z_0$	$= Z_0$
Component to be measured	E and H	E or H	E or H

While spatial EMF components in the near-field are calculated, the rigorous use of the general dependencies [for instance, Eqs. (2.3) and (2.4)] is indispensable and appropriate precautions should be taken when any simplifications in calculations are planned. Special caution is necessary when applying software for numerical analysis without an appropriate analysis of the simplifications and assumptions that have been accepted in the procedures.

As noted in Section 2.1, properties of EMF in the far field appear partly in the intermediate field as well, although none of them appears in the near field. This results in the necessity of the specific methods use for EMF measurements in both regions. Several examples are quoted below to illustrate this point:

- In the far field, E and H measurements are fully equivalent, and they permit the calculation of the other components as well as S. In the near field, separate E and H component measurements are indispensable, and they significantly complicate the issue of the S measurement.

- The EMF polarization in the near field, especially in conditions of multipath propagation, may be quasi-ellipsoidal because of the spatial orientation variations of the polarization ellipse. This is due to, for instance, the frequency of source variations as a result of its FM modulation, Doppler effect due to reflection from a moving object, and so on.

- The radiation pattern in the far field is constant and independent of the distance to a source; on the ground of the near-field measurements, it may be calculated only for sources of regular structure using complex computations [5].

- The Poynting vector in the near field is complex, and its direction and magnitude are functions of the source structure and the distance to the source.

A comparison of the requirements in near- and far-field EMF measurements is presented in Table 2.2.

2.3 EMF FROM SIMPLE RADIATING STRUCTURES

If in Eqs. (2.3) and (2.4) we assume that the electric current has a non-zero magnitude in the direction of axis z, i.e.: $*J = 0$ and $|J| = J_z = $ constant, for:

$$-1/2 < z < +1 \,/2$$

and at the same time

$$l << \lambda \text{ and } l << R$$

(where l = the length of the dipole's arm), then the calculated Π is substituted into Eqs. (2.1) and (2.2) to give the components of the EMF generated by an electric elemental dipole placed in the Cartesian coordinate system as shown in Fig. 2.4.

The components are:

$$E_R = \frac{p}{2\pi\varepsilon} \left(\frac{1}{R^3} - \frac{jk}{R^2} \right) \cos\theta \, e^{-jkR}$$

$$(2.21)$$

Table 2.2 Specificity of the Near-Field EMF Measurements

Parameter	Near field	Far field
Measured EMF component	E, H, and S	E or H, and S on micro-waves
Other magnitudes measurement	I, T, (SA, SAR) "HESTIA"	Unnecessary
Spatial components	3	1 or 2
Polarization	Quasi-ellipsoidal	Linear or elliptical
Environment	Complex, multipath propagation, and interference	Usually simple
Frequency spectrum	Wide, often unknown, many fringes	Usually single frequency
Antennas	Small, omnidirectional	Resonant, directional
Temporal and spatial EMF alternations	Significant	Usually negligible
Uncertainty	3, 6 or more dB	Around 1 dB
Temperature sensitivity	Significant	Unessential
Susceptibility	Significant	Omittable
Influence of surroundings	Significant	Usually omittable
Procedures	Complex	Simple
Agreement with theory	Reasonable	Good
Measured levels	V/m, kV/m	mV/m, µV/m, dBµV/m

$$E_\theta = \frac{p}{4\pi\varepsilon} \left(\frac{1}{R^3} - \frac{jk}{R^2} - \frac{k^2}{R} \right) \sin\theta \, e^{-jkR}$$

$$(2.22)$$

$$H_\varphi = \frac{j\omega p}{4\pi} \left(\frac{1}{R^2} - \frac{jk}{R} \right) \sin\theta \, e^{-jkR}$$

$$(2.23)$$

where p = the dipole moment:

$$p = \left| \mathbf{p} \right| = \left| \frac{I_z \, dz}{j} \mathbf{1}_z \right|$$

$$(2.24)$$

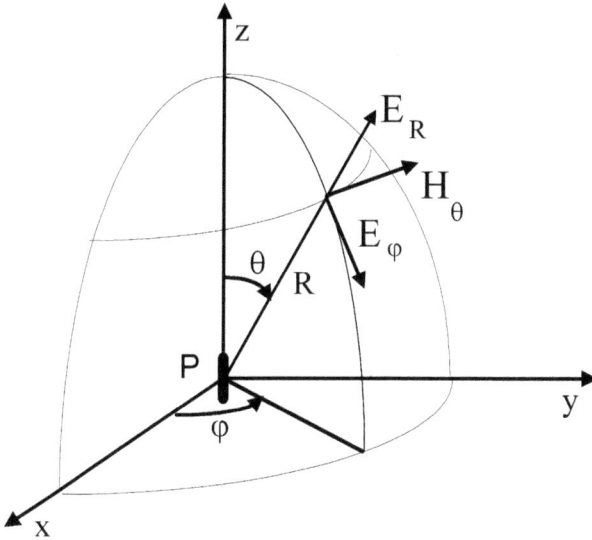

Figure 2.4 Elemental electric dipole in Cartesian coordinates.

and

I_z = the current in the dipole,

$\mathbf{1_z}$ = the versor (unit vector) of axis z.

If we repeat similar procedures for $J = 0$ and $|\overset{*}{\mathbf{J}}| = \overset{*}{J_z}$ = constant, we will have formulas defining the spatial field components of the elemental magnetic dipole. Using the principle of duality, the formulas may be immediately obtained from Eq. (2.21), (2.22), and (2.23) by the exchange of E to H and ε to μ and conversely.

The maximal spatial variations of the EMF components, as a function of distance to a point of observation, may be written in the following forms:

- For an elemental electric dipole:

$$E = C_1\,(\alpha)\,\frac{\exp\,(-jkR)}{R^\alpha}$$

$$\tag{2.25}$$

while

$$3 \geq \alpha \geq 0 \tag{2.26}$$

for: $0 < R < \infty$

where $C_1 (\alpha)$ = a constant dependent of R, and E = $|\mathbf{E}|$ is the complex amplitude of the electric field strength.

- And by analogy:

For an elemental magnetic dipole:

$$H = C_2 (\alpha) \frac{\exp (-jkR)}{R^{\alpha}}$$

(2.27)

where $C_2 (\alpha)$—a constant dependent of R, and H = $|\mathbf{H}|$ is the complex amplitude of the magnetic field strength, while α and R fulfill the conditions of Eq. (2.26).

Equations (2.25) and (2.27) illustrate the curvature of the EMF and are not related to any particular field zone. The radius of curvature is proportional to $R^{\alpha+2}$ and varies from infinity (for the plane wave) to very small magnitudes in the near field.

If we substitute $|\mathbf{J}| = J_z \sin k(h - |z|)$ and $*\mathbf{J} = 0$ in Eqs. (2.3) and (2.4) within the limits $- h \leq z \leq + h$ (where h is the length of the dipole arm), after calculation of Π using Eqs. (2.1) and (2.2), we will find components of the EMF from an infinitely thin, symmetrical dipole antenna of total length 2h (Fig. 2.5). As a result of the sinusoidal current distribution assumption in the dipole, unlike the "ideal" assumption of a constant current distribution, a certain error is expected. The error is especially important for infinitesimally thin (one-dimensional) dipoles. However, the assumption is fully acceptable while the radiation pattern of such antenna is being considered. The use of precise solutions of the integral equations, describing current distribution in a real antenna, is

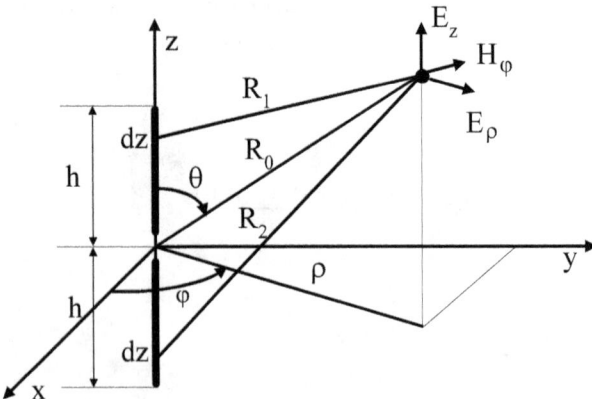

Figure 2.5 Symmetrical dipole in coordinate system.

not necessary in the aspect of the EMF components' strength calculations in the area of interest as well as in the light of the final conclusion of the considerations presented [6]. Here we will use the results of calculations available in the literature [1]. The EMF components E_z, E, and E are given by Eqs. (2.28), (2.29), and (2.30) respectively:

$$E_z = -\frac{jZI_0}{4\pi \sin kh}\left[\frac{\exp(-jkR_1)}{R_1} + \frac{\exp(-jkR_2)}{R_2} - 2\cos kh \; \frac{\exp(-jkR_0)}{R_0}\right]$$

$$(2.28)$$

$$E_\rho = \frac{jZI_0}{4\pi\rho \sin kh}\left[\frac{\exp(-jkR_1)}{R_1}(z-h) + \frac{\exp(-jkR_2)}{R_2}(z+h) - 2\cos kh \; \frac{\exp(-jkR_0)}{R_0}\right]$$

$$(2.29)$$

$$II_\phi - \frac{jI_0}{4\pi\rho \sin kh}\left[\exp(-jkR_1) + \exp(-jkR_2) - 2\cos kh \exp(-jkR_0)\right]$$

$$(2.30)$$

where I_0 = current intensity at the dipole input,

$$R_1 = \sqrt{\rho^2 + \left(z - h\right)^2}$$

$$R_2 = \sqrt{\rho^2 + \left(z + h\right)^2}$$

$$R_0 = \sqrt{\rho^2 + z^2}$$

and other indications as in Fig. 2.5.

If we continue here considerations similar to the above, in the case of the electric and magnetic elemental dipoles, and generalizing on the basis of Eqs. (2.28–2.30), per analogy to Eqs. (2.25) and (2.27), we can write the relations describing EMF variations in the proximity of the symmetrical dipole:

$$E = C_3 (\alpha) \frac{\exp(-jkR)}{R^\alpha}$$

(2.31)

$$H = C_4 (\alpha) \frac{\exp(-jkR)}{R^\alpha}$$

(2.32)

where:

$$1 \geq \alpha \geq 0$$

$C_3(\alpha)$ and $C_4(\alpha)$ = constants dependent of R,

R = the distance to a point of observation;

while for EMF close to the antenna surface, one may assume $R \approx \rho$.

If in Eqs. (2.25) and (2.27) we substitute $\alpha = 3$, we will have a relationship defining the EMF variations as a function of distance in the near field of elemental dipoles where $\alpha = 2$ represents the intermediate-field of the dipoles. The far field of the elemental dipoles and the near field of a thin symmetrical dipole antenna are characterized by $\alpha = 1$. The variability of the later versus distance is specific to a spherical wave. Rigorous analysis of Eqs. (2.21–2.23) and (2.28–2.30) does not justify an assumption in the formulas that $\alpha = 0$ for $R \Rightarrow \infty$, which would represent the plane wave. Such a simplification is often accepted when an EMF in a limited area, sufficiently far from a source, is being considered. In that area, amplitude variations of E and H vectors in any direction are negligibly small. The simplification is equivalent to the assumption that the radius of curvature of the field considered is equal to infinity. The maximal phase variations are independent on α if one assumes $\alpha = $ const; such a case is most interesting from the point of view of metrological practice.

The comparison of Eqs. (2.25) and (2.27) as well as (2.31) and (2.32) permits us to come to the conclusion that the EMF "variability" in proximity to sources much smaller in the comparison to the wavelength ($\alpha = 3$) is the largest. Thus, if we estimate the errors of the EMF measurements near the sources, we will have majorants of the errors for an arbitrary source. The conclusion is, in some sense, an intuitive one, and it is a result of the presence of the quasi-stationary field in proximity to sources whose sizes are comparable or larger than the wavelength (induction field). A particular example of the case is the EMF in the proximity of AC power devices and especially around overhead transmission lines. Usually, only Coulomb's and Biot-Savart's laws are applied, and this approach is equivalent to the assumption that the

EMF does not exist, and the field is sufficiently represented by E and H fields only.

Doubt may arise, under these considerations, relating to the presence of higher powers of α when a multipole expansion is applied. The approach may make more precise calculations of EMF generated by elemental sources more possible. However, even if appropriate corrections are applied, it does not make a substantial difference to the maximum errors. This is especially true while arbitrarily small-sized physical sources are considered. It should be emphasized here that only physical sources have a practical importance because of the efficiency of the EM energy radiation. Good examples here are formulas describing the standard EMF near, for instance, a standard loop antenna. In this case, apart from the finite sizes of the antenna that are much larger than the elemental dipole, no term exceeds a power of 3. This is a matter of practical importance [7]. Having analyzed the near- and far-field terms, the next chapter considers methods of measurement.

References

[1] D. J. Bem. *Antennas and radiowave propagation* (in Polish). Warsaw 1975.

[2] B. Minin. *VHF radiation and the human security* (in Russian). Moscow, Sovietskoe Radio 1974.

[3] Ju. D. Dumanskij, A. M. Serbyuk, I. P. Los. *The influence of RF electromagnetic fields on humans* (in Russian). Kiev 1975.

[4] FprEN 50492:2008 Basic standard for the in-situ measurement of electromagnetic field strength related to human exposure in the vicinity of base stations.

[5] P. F. Wacker. *Non-planar near-field measurements: spherical scanning*. Natl. Bureau of Standards Publ. NBSIR 75–809, Boulder, CO USA.

[6] A. Karwowski, P. Buda. *The method of the protection zones in the proximity of medium- and long wave transmitting antennas* (in Polish). Prace IL No. 93/87, pp. 2–27.

[7] H. Trzaska. *Magnetic field standard at frequencies above 30 MHz.* HEW Publications, (FDA) 77–8010, vol.II, pp. 68–82, Rockville, MD.

Chapter 3

EMF Measurement Methods

In order to select an optimal method for EMF measurement in the near field, it is first necessary to determine which quantities best characterize the field. These quantities will then be the subject of the measurement.

From the point of view of antenna performance evaluation, it is essential to measure the strength of E or H components near the antenna. This then makes it possible to find the current or the charge distribution along the antenna. With this as a basis, it is possible to find the radiation pattern of the antenna and its input impedance. The measurement of E, H, or S in the near field (with the phase information conserved) permits us, with some complex calculations, to find the antenna's radiation pattern in the far field. From the point of view of shielding, absorbing, or EMF attenuating materials investigations, the E, H, and S measurements are sufficient as well.

If we are interested in protection against unwanted exposure to EMF, and in biomedical investigations in particular, the E, H, and S measurements are not sufficient. This area of investigation requires more precise qualification of the parameters that are specific for these purposes, and these should be a subject of the measurement. The proposals cited previously for the protection standards provide, as the basic criterion of the interaction of the EMF with biological media, the power or energy absorbed in the mass unit [specific absorption rate (SAR) and specific absorption (SA); see Table 1.1]. Sometimes the absorbed power in volume units is applied, and a widely accepted measurement is the temperature rise due to EM energy absorption (which permits the determination of SAR and SA). The measurement of current induced in a body by external EMF has recently become more popular. The majority of protection standards require the measurement of the root-mean-square (RMS) value that reflects the quantity of the absorbed energy. In the nonthermal approach,[*] it is more important to know the amplitudes of the field components, their spatial positioning, and their temporal variations, as well as the frequency of the carrier wave and that of the modulating spectra (and their temporal variations) and the type of modulation (AM, FM, PM).

3.1 E, H, and S Measurement

In Chapter 1, portions of several versions of protection standards were presented to illustrate the range of measured magnitudes of E, H, and S. It should be repeated that these magnitudes show metrological requirements only for the surveying and monitoring services. Only laboratory experiments will require field measurements from the lowest measurable magnitudes (near the noise level or even below the noise level) to the highest, which can be generated by the use of available power sources. Moreover, the levels given by the standards vary in the succeeding versions, modifications, and implementations of those standards. The other parameters of the measured field are much less rigorously defined in the standards. It is helpful to investigate those parameters here.

3.1.1 Frequency (Spectrum) of the Measured EMF

At an arbitrary moment of time, in a chosen point in space, there exists a solitary vector E and a solitary vector H. They are linearly polarized, and their magnitude is equal to the sum of instantaneous values of any spatial components and spectral fringes appearing at the point considered. The condition may be written in the form:

$$E = E_0 + \sum_{i=1}^{N} E_i \cos(\omega_i t + \varphi_i)$$

$$(3.1)$$

where:

E_0 = the electrostatic field strength,

E_i = the strength of the i-th spectral component,

ω_i = the angular frequency of the i-th component,

---

[*] Although presently unfashionable, in the authors' opinion, the future belongs to this approach.

φ_i = the phase of i-th component,

N = the number of field components to be considered.

If we substitute H for E in Eq. (3.1), we will obtain the formula defining the temporal variations of the magnetic field. If we neglect the static component in the formula, we can note that, without regard to the region considered (Fresnel or Fraunhofer zone), and with the exception of guided waves, E is orthogonal to H. We should note here that the spatial positioning of the resultant vector is not given by the formula, and the positioning may be arbitrary one. The sum given by Eq. (3.1) is a finite one. In many cases of practical importance, N does not exceed 1 or 2. However, even in the simplest cases, simultaneous measurement of all the frequency components may be technically difficult or even impossible as, for instance, in the case of simultaneous measurement of static and RF components. At times, it may be undesirable because of interpretational problems—for instance, when the measurements are performed in conditions where the components fall in frequency ranges where different levels are permitted.

The issue has four important aspects:

1. It is technically possible to construct an EMF meter with a frequency response equivalent to the frequency dependencies of the limits determined by the protection standards. While the frequency response of a meter is a continuous function of frequency, the protection standards, as well as any their modifications and new proposals, are characterized by discontinuities at the borders between frequency ranges where the limits are different. The meters are compatible with one standard and do not permit evaluation of the actual roles of the separate EMF sources, especially when they work within different (in the aspect of protection standards) frequency ranges.

2. Wideband EMF measurement by inspection services seems to be the most convenient technique because of the speed and simplicity of the measurement. However, in order to assure unequivocal results of the measurement, the use of a meter covering more than one frequency range, as represented in the standards, requires switching off any other source apart from the measured one. However, even in such a situation, the presence of harmonics in the spectrum radiated by the source may lead to problems with interpretation of the results of such a measurement.

3. The issue is of primary importance in the near field; even in the proximity of telecommunication devices and systems, which should

radiate only in a channel (-s) and spurious radiations are limited by radiocommunications regulations. Near them appear harmonics of the basic frequency, intermediate frequencies, and their combinations. In this case, only selective measurements may reflect the whole radiated spectrum.

4. A selective measurement can also be troublesome, especially when measurements are performed in the presence of a larger number of sources. However, the measurement allows precise estimation of the role of any separate EMF source in the resultant field. A new concept of such a measurement is presented in Chapter 9. The authors began their involvement in the field from the selective meters designed in the early 1960s. In the authors' opinion, the selective methods may be considered as the best methods for the future.

3.1.2 EMF Polarization

The expression *polarization* is understood in three ways:

1. As positioning of the vector E in relation to a chosen reference system; e.g., in relation to the Earth's surface (vertical and horizontal polarization).

2. As the shape of an envelope of the E (or H) vector rotations in space (linear polarization, circular polarization, elliptical, quasi-ellipsoidal).

3. As the direction of the E (or H) vector rotations in the space (left- and right-hand polarization).

We will only, generally, concern ourselves with the first two meanings of *polarization*.

The maximal value of E and H does not result from polarization in any of the above senses, whereas the RMS value depends on the polarization only in the sense defined in item 2 above—where the magnitude of the energy absorbed by a body, for instance by a human body, as well as the current induced by an EMF in the body are a function of the field vectors' position in relation to the body. In the latter case, polarization in the senses described in items 1 and 2 is of concern. This shows the importance of the EMF polarization in our considerations and the necessity to be able to measure this parameter of EMF as well.

It is necessary to call attention here to the dependence of the results of EMF measurements on the polarization of the measured field and the directional pattern of the probe applied. We must understand the advantages and disadvantages of probes with sinusoidal, circular, and spherical directional patterns when an EMF of an unknown polarization is being measured. The problem will be briefly discussed in Chapter 7.

3.1.3 EMF Modulation

Every single quantity represented in Eq. (3.1) may be the subject of intended variations as a function of time (modulation). The variation of E is called *amplitude modulation* (AM), and one of the very important types of modulation (extremely essential in considered aspect) is *pulse modulation*. When the subject of variations is ω, we call it *frequency modulation* (FM), and in case of φ alternations, it is called *phase modulation* (PM).

When monochromatic fields are being measured, there are almost no problems with the interpretation of the results. In the case of modulated field measurement, and in particular when pulsed fields are measured, the question arises as to what we should measure: maximal instantaneous value, mean value, or RMS? The answer to the question (doubt) should be provided by biologists and medical doctors based on detailed studies of the importance of thermal interactions (RMS measurement) or nonthermal ones (peak value measurement). The role of an engineer should be an auxiliary one, as a supporting consultant during basic laboratory studies and as the person responsible for making appropriate choices for measuring devices or an exposure system fulfilling requirements of the experiment to be carried out.

Let's focus our attention on a technical aspect of the problem. In further considerations, it will be shown that the RMS value is measured by a probe using a square-law detector. The indication of the meter is proportional (equivalent) to the RMS value of the sum of any spectral fringe in a particular frequency band. However, the design of a correctly functioning square-law detector, especially when the probe is intended to work in the near field, in a wide frequency range, and with a large dynamic range, is difficult, and such a probe has yet to be constructed. The measurement of the instantaneous peak value, especially of short monopulses, is extremely troublesome, and its realization requires the use of expensive measuring devices and complex analytical methods to reconstruct the shape of the measured pulse. It is possible that the standards should suggest (or even require) the simultaneous measurement of both values; however, that would increase the costs of the measurements and make them more burdensome. In order to simplify the measuring procedures, as well as to decrease costs of the typically expensive measuring equipment, the majority of the meters available on the market are provided for measurements of the EMF of an uninterrupted envelope, although the limitation is rarely mentioned in the manuals of these devices.

The measurement of E, H, and S is usually performed with the use of probes based upon an electric or magnetic antenna of small electrical sizes and loaded with a diode detector. Detailed considerations related to these probes are presented in Chapter 4 and the following. Magnetostatic fields or very low frequency (VLF) magnetic fields are usually measured with the use of Hall-effect devices or other types of semiconducting devices. Although a detailed analysis of them is not taken into account, some of the considerations presented here may be helpful when these sensors are applied.

3.1.4   The Use of Far-Field Meters for Near-Field Measurements

The basic features of near-field EMF measurement devices is the small size (both in the physical and electrical sense) of their measuring probes as well as their possibly poor directional properties. A bit less evident is the necessity of using electric field sensors based on electric antennas (whip, symmetric dipole antenna) and sensors with magnetic antennas (loop, ferrite rod) for magnetic field measurement. An exception to this rule will be discussed in Chapter 6, as confusion may arise from the widespread use of the meters equipped with loop antennas (because of their better stability, reduced sensitivity to the presence of conducting objects in their proximity, and relatively smaller sizes) and calibrated in E-field units. *The latter may be used for far-field measurements only,* where the constant relation between electric and magnetic field exists, as shown by Eqs. (2.13) and (2.14). The phenomenon (as evident) will not be discussed in subsequent parts of this work. It requires, however, a few words of comment, as even people experienced in EMF measurements often make such a mistake (gross error), and results of such measurements may be found in the literature. The power density S is also often measured by the way of E or H measurements. While it is very appropriate in the far field, knowledge and caution are required when such a measurement is performed in the near field. It is hoped that it is not necessary to add here that power density meters, equipped with resonant-size antennas (horn or log-periodic antennas), are completely useless in near-field measurements. We shall recall here the far-field boundary estimations presented in the previous chapter: the far-field limit of a relatively small-sized parabolic antenna may exceed several hundred meters, or even more! Not to mention indoor measurement using these antennas (again, the results of such measurements are available in the literature).

In the near field, the mutual relationship between both field components is unknown *a priori*. The relationship depends upon the structure (type) of the source of radiation, and it is a function of direction to the source and distance between the source and the point of observation. The exception to the rule, expressed by Eqs. (2.13) and (2.14) and valid only in the far field, is sometimes used as one of the possible criteria for the far-field boundary [1].

Criteria for evaluation of the measuring antennas' sizes and the directional pattern of the probes used are subject of detailed analysis presented in later chapters. They are mentioned here only to focus our attention on the most important features of the meters used in near-field metrology.

3.2 TEMPERATURE RISE MEASUREMENTS

In order to illustrate the energy transfer from the EMF to a phantom, let's consider the simplest case of a lossy dielectric immersed between plates of a capacitor, as shown in Fig. 3.1.

Between two identical plates of surface area S, and placed at distance d, is an immersed lossy medium that occupies the whole volume between the plates. Voltage V is applied to the plates. This situation is typical for effectiveness analysis of dielectric heating. The power P absorbed in the medium is:

$$P = VI\cos\phi \approx VItg\delta$$

(3.2)

where:

I = current flowing through the medium:

$$I = V\omega C$$

(3.3)

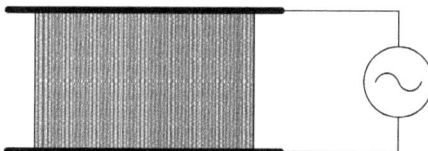

Figure 3.1   A lossy medium between two plates.

φ = phase shift between V and I,

δ = dielectric loss angle:

$$\delta = 90° - \phi \tag{3.4}$$

ω = angular frequency,

C = capacitance of the set:

$$C = \varepsilon_0 \varepsilon_r \frac{S}{d} \tag{3.5}$$

ε_r = relative permittivity of the medium.

We obtain the energy, W, absorbed in the medium by multiplying the power by time of exposure t:

$$W = Pt = \omega \varepsilon_0 \varepsilon_r vtE^2 tg\delta \tag{3.6}$$

where:

E = electric field intensity between the plates:

$$E = V / d \tag{3.7a}$$

v = volume of the medium, v = Sd.

The energy, W, is transformed into heat, Q:

$$Q = cv\rho\Delta T \tag{3.7b}$$

where:

c = specific heat of the medium,

ρ = mass density,

ΔT = temperature increase.

Without taking into account the heat transfer to the surroundings (by radiation or conduction)—i.e., taking into account only the thermal capacity of the body, in the conditions of full thermal insulation—the energy, W, must be equal to the heat, Q. Thus, if we compare Eqs. (3.6) and (3.7) and then we calculate t (time necessary to warm the body in ΔT), we will have:

$$t = \frac{\rho \, C_p \, \Delta T}{\varepsilon_o \, \varepsilon_r \, \omega \, E^2 \, tg \, \delta}$$

(3.8)

If in Eq. (3.7) we substitute the mean magnitudes of the living tissue parameters and we assume the minimal measurable increase of temperature $\Delta T \approx 0.1$ K and E = 10 V/m, then for such idealized conditions (without taking into consideration the heat transfer!), the time required for the temperature rise (in 0.1 K) is 10^5 to 10^{10} s, depending upon frequency of the field. The heating time versus frequency, for E = 10 and 100 V/m, is plotted in Fig. 3.2. This is the most important factor limiting the method application in EMF measurements—its sensitivity.

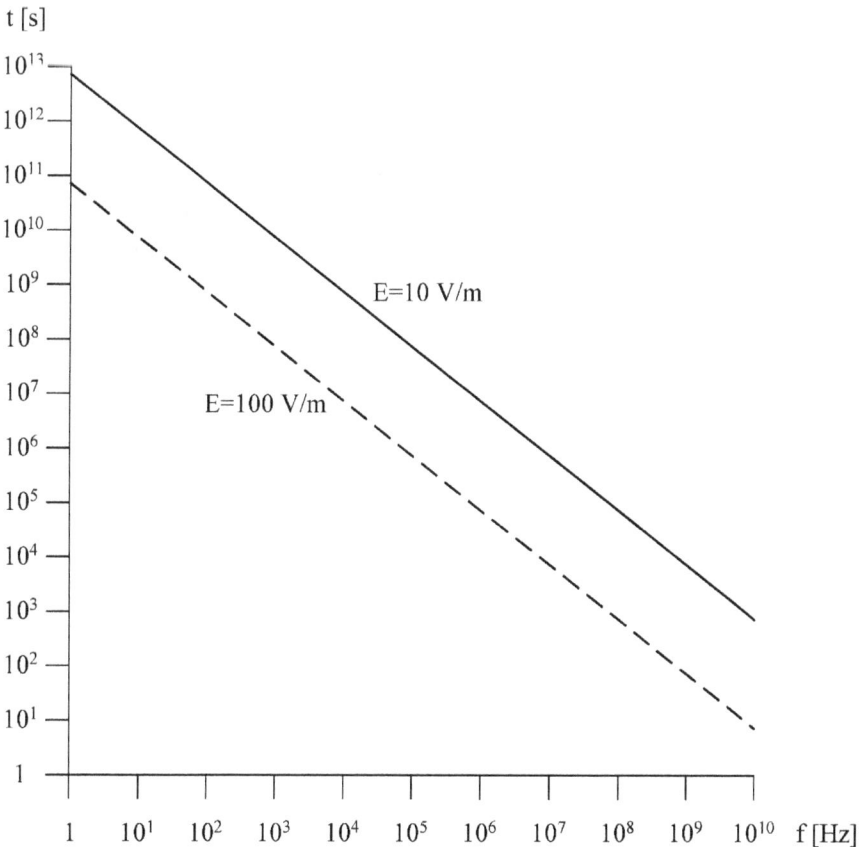

Figure 3.2   Heating time as a function of frequency for ΔT = 0.1 K.

While we accept any without question the advantages of the SAR as the quantity applied in the basic research, let's analyze its applicability in metrological practice.

The SAR is defined twofold:

1. As a measure of the temperature increase of an exposed object:

$$SAR = \frac{C_p \Delta T}{t} \left[\frac{W}{kg} \right]$$

(3.9)

2. As a measure of EMF energy absorption:

$$SAR = \frac{\sigma E^2}{\rho} \left[\frac{W}{kg} \right]$$

(3.10)

where:

σ = conductivity of the medium,

ρ = mass density.

(Note here that E is the rms value, so it is equivalent to other equations in the literature that take the magnitude of the vector E and then have a constant of value 2 in the denominator.) Both SAR definitions should be equivalent. Let's make a use of the equivalence and to compare them:

$$\frac{\sigma E^2}{\rho} = \frac{c \Delta T}{t}$$

(3.11)

Now we will calculate t:

$$t = \frac{c \rho \Delta T}{\sigma E^2}$$

(3.12)

And making evident substitution:

$$\sigma = \varepsilon_0 \varepsilon_r \omega tg\delta$$

(3.13)

Equation (3.12) becomes identical to Eq. (3.8). It allows the conclusion that the definition of the SAR was introduced in a similar mechanistic way as the above example.

The temperature rise measurement and then SAR calculation may be characterized as follows:

- The temperature rise represents the best quantity of EMF energy absorbed by a body; a good agreement between theoretical analyses and experimentation is obtained without regard to the field modulation, polarization, etc.

- Because of thermoregulation mechanisms *in vivo*, the relation between the measurements *in vitro* and phenomena *in vivo* creates some doubts; similar doubts arise due to the heat transfer from the body to its surroundings.

- Exposure measurement using a phantom permits preservation of the full analogy between the conditions of measurement in relation to an exposed person. (The analogy does not exist in any other measuring method—the large size of the "probe," one of the most important inconveniences of measuring probes, here may be considered an advantage.)

- The use of such a "meter," especially for practical purposes, seems to be extremely troublesome.

- Measurement repeatability can be difficult to obtain. Measurements performed *in vivo*, although possible, seem to be burdened with a substantial measuring error resulting from the difficulty in differentiating bioeffects caused by an exposure and those due to necessary damage of a living body caused by the placement of a sensor within it.

- There is no realistic possibility of distinguishing the polarization of the EMF illuminating the absorber (phantom) or its frequency; because of the thermal inertia, it is not possible to measure the EMF modulation.

- The frequency response of the phantom is a function of its dimensions and shape as well as a function of the position of the phantom in relation to EMF vectors.

- As mentioned above, the sensitivity of the method is insufficient.

- A doubt appears here as to whether and how to measure the average value of the absorbed energy for the whole body (phantom) or a point value.

Temperature measurement methods are well known. However, for clarity, some selected methods of temperature measurement are presented below.

3.2.1 Temperature Measurement with the Use of Liquid Crystals

The essence of this method is based on the investigation of the tincture or the light reflection coefficient of a liquid crystal. The crystal is immersed in a micro container and illuminated with the use of an optical fiber. Another fiber leads light reflected from the surface of the crystal to a photodetector.

Here it is possible to achieve resolution of 0.1 K under conditions of frequent calibration of the sensor. Without the calibration, because of thermal drift and ageing of the crystal, the resolution decreases to about 0.25 K.

Because of the absence of a conducting component in the device's design, the sensor (including feeders) is "transparent" to the measured field, which eliminates the measured field disturbances caused by it and, as a result, increases accuracy [2].

3.2.2 Temperature Measurement with the Use of a Thermoelement

The method is based un the use of a thermoelement immersed in a thin-walled glass pipe, which is then inserted into the tissue under investigation. To limit disturbances in the measured field by the metal leads of the thermoelement, as well as to eliminate the possibility of the EMF penetration into the tissue by this path, the measurement is performed before and after the exposure of the tissue. While it is being exposed, the thermoelement is withdrawn from the pipe [3].

3.2.3 Thermistor Temperature Measurement

The use of a thermistor inserted into tissue allows continuous observation of its temperature variations, including while the tissue is exposed to the EMF. To limit measurement errors the resistance variations are measured by a bridge.

This method was modified to increase sensitivity and measurement accuracy [4]. For these purposes, a high-resistance thermistor was applied as well as transparent leads of 160 kΩ/cm resistivity. The latter carry a 0.3-μA DC exciting current. An additional pair of leads, connected directly to the thermistor, permits measurement of the voltage drop across the thermistor. The DC power dissipated by the thermistor does not exceed 0.1 μW, while that in the leads is below 0.05 μW/cm. The resolution of the device exceeds 0.01 K.

3.2.4   Temperature Measurement with the Use of a Viscosimeter

Variations of a liquid viscosity as a function of temperature allow the use of this phenomenon for temperature measurements. The presented device [5] includes a system of capillary tubes throughout which a liquid is pumped. The pressure difference in a capillary at the input of the sensor and at its output is a measure of the temperature. The pressure difference is measured by a transducer. The measuring ranges and the sensitivity of the device changes are achieved via the choice of liquid.

3.2.5   Thermographic and Thermovisional Measurements

The development of theoretical analyses of absorption models has led to "millimeter resolution" models [6]. Although the models are not the subject of the work, we should recall that the first model studies, initiated by Guy and Johnson, were followed by experimental studies that made it possible to verify the theory and the correctness of the necessary simplifying assumptions in it. The measurements were performed by applying thermovision and thermograph cameras. The experimental models, of different geometric shapes and sizes, contained several parts that permitted observation of different cross sections of the model and, as a result, the temperature distribution in the sections after the model exposure. The models were usually electrically homogeneous and isotropic. However, they allowed the researchers to obtain many interesting results showing the dependence among shapes, sizes, and electrical properties of the model (phantom) on one hand and the manner of exposure, frequency, EMF polarization in relation to the object, and modulation on the other [7]. Experimental results can be expensive to obtain and, because of good agreement between models and measurements, some experiments may not need to be undertaken.

These methods have allowed quick, simple, and easy visualization of the temperature distribution in a chosen plane of the model and, as a result, localization of the thermal extrema (hot spots) while different combinations of exposure are used. A disadvantage of the method is its thermal inertia and the necessity to have an approach to (visibility of) the investigated area. On the other hand, its doubtless advantage is the possibility of a noninvasive measurement at a distance, with no physical contact between a sensor and a body (remote sensing), which permits us to limit EMF disturbances in the proximity of the exposed body as well as direct use of the measurement results for archiving and computer analysis.

3.3 SAR MEASUREMENT

The above-presented discussion of temperature measurement was mainly devoted to SAR measurement on the basis of the SAR definition given by Eq. (3.9). Contrary to a "thermal" definition of the SAR, the definition given by Eq. (3.10) is electrical in nature. It is irreplaceable in numerical dosimetry. However, the use of the definition for measuring purposes is loaded with significant errors resulting from the accepted method. The SAR measurement concept is based on E-field measurement. An evident solution here would be an anthropoidal phantom, filled with a liquid of parameters similar to that of human tissues, and E-field measurement would be performed using an E-field probe. The solution is fully acceptable; however, its application is limited to laboratory conditions only. In order to approach the measuring conditions of SAR probes (E-field ones) designated for measurements outside the lab, the E-field probe is immersed in a sphere filled with a medium whose parameters mimic those of a living body (Fig. 3.3).

Causes of errors in the approach include

1. Different propagation properties of a small sphere in relation to dimensions of a body and different frequency dependence of absorption (resonant frequencies). The absorption is a function of polariza-

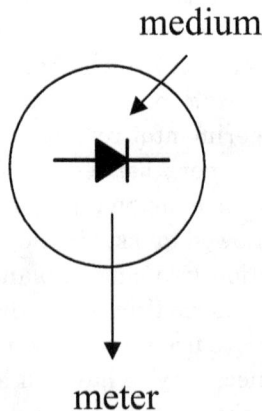

Figure 3.3 Concept of the "SAR probe."

tion (and it is different for different spatial EMF components), reflection, refraction, wave type (plane wave, spherical, or cylindrical), and wave zone (induction, near, or far field).

2. The choice of the electrical parameters of the medium is almost an arbitrary decision, and it does not reflect the complex structure of a living body; different parameters of the same organs and tissues in different people; different shapes, weights, metabolisms, thermoregulations; or even lifestyle of the person, all of which should be represented by the probe.

Conclusion: an estimation of the equivalence error of the SAR measured this way and the real magnitude of the SAR in a real body shows values on the level of one order of magnitude or worse.

3.4 CURRENT MEASUREMENTS

Unlike temperature measurements, measurement of the current, induced in a human body by EMF is already the subject of regulations. The measurement is especially useful as a measure of the hazard created by EMF at the lowest frequency ranges. Sources of such fields include the neighborhood of overhead high-voltage transmission lines, power substations, and close to long- and medium-wave high-power broadcast transmitters, where polarization parameters of the measured EMF are well known. An additional advantage of the measurement (and especially corresponding to legal regulations) is the possibility of including EM radiation hazards and electric shock in one protection standard.

The current measurement is undertaken in one of three ways: (1) by placing a human subject on a conducting, standard-size plate and measuring the current between the plate and the surface (of the earth) using a thermocouple, (2) by measuring the voltage drop on a resistance between the plate and the ground, or (3) by using a current transformer (Fig. 3.4).

In contrast to an electric shock, the use of the current measurement in the case of an EM radiation hazard has some causes of inaccuracies, including:

- The result of the measurement depends not only on the posture of the people being measured but also on the clothes they are wearing, particularly their shoes.

- The current measurement accuracy is extremely unsatisfactory.

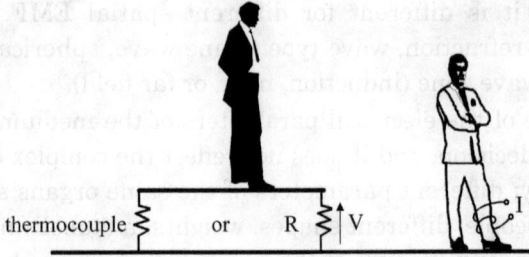

Figure 3.4   Methods of measurement of the current flowing through a human body.

- Current measurement, in the manner shown in Fig. 3.4, in accordance with standards in effect, entirely reflects currents induced in the body by EMF components parallel to the vertical axis of the standing person. It is impossible to measure horizontal components of the current, excluding currents induced in the body by H-fields (eddy currents). The latter have been the subject of intense biomedical investigations lately.

Although these measurement methods are concerned only with current measurements in the foot or leg, it has also been shown that large currents may flow throughout other parts of the body, for example, a current in the hand or lips of a person using a walkie-talkie [8]. Two series of experiments were performed. The first involved the use of a model of an operator's hand, as shown in Figs. 3.5 and 3.6 [9]. The model contains a 1.5-m long glass tube of cross section 60 × 110 mm filled with 5 percent saline solution, the level of which (the hand's length) may be changed using a discharge valve in the bottom part. The bottom of the tube is covered by a metallic plate that is connected through a thermocouple with a transceiver.

During the experiments, different lengths of the "hand" were applied, and current measurements were performed for different placements of the transceiver's antenna in relation to the thermocouple input. The measurements were performed with the use of standard radiotelephones with an output power of 5 W. The power was not measured but was taken from the devices' manuals. This calls attention to the problem of power measurement. The problem was neglected in all of our experiments, although, due to the varying size of the radiating system, antenna matching and radiated power could be reduced. Selected results of the measurements and estimations are shown in Table 3.1 for different antennas, frequencies, and hand lengths (1).

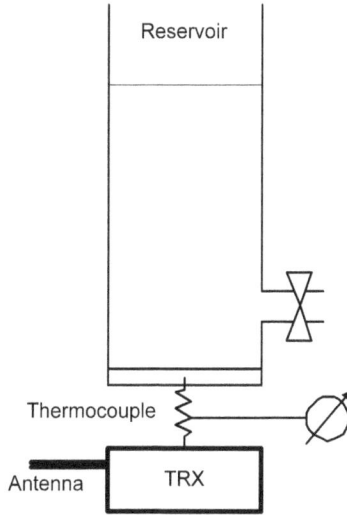

Figure 3.5 A hand's model.

Figure 3.6 View of the hand's model during measurements.

Table 3.1 Selected Results of the Hand's Current Estimations and Measurements

	f = 150 MHz antenna λ/4 1 = 1100 mm side fed	f = 150 MHz antenna λ/4 1= 1100 mm end fed	f = 150 MHz antenna λ/4 1= 550 mm end fed	f = 430 MHz antenna λ/4 1= 550 mm side fed
Theoretical	61.4 mA	59.3 mA	59.3 mA	188.0 mA
Measured	75 mA	69 mA	65 mA	58 mA

The results of hand's and lips' current (I_h, I_l) measurements, measured *in vivo* for several types of hand-held radiotelephones of 5-W nominal output power and supplied with different antenna types, are shown in Table 3.2. The column ΔE shows the increase in E-field strength, measured at distance of 10 m from the device, while the radiotelephone is held in a hand by its operator in relation to the same device placed on an insulating support. The results shown in the table emphasize the role of an operator's body as a "counterpoise," especially at lover frequencies.

Table 3.2 Hand and Lip Currents of a Radiotelephone Operator

Frequency	ΔE	I_h [mA]	I_l [mA]	Antenna	Power [W]
27 MHz	15×	170	120	1.5 m	5
27 MHz	10×	150	100	25 cm	5
144 MHz	3×	90	70	5/8 λ	5
144 MHz	2×	80	70	15 cm	5
432 MHz	1.5×	50	50	12 cm	5

This may be confirmed by the presence of the standing waves on the arm of an radiotelephone operator and the E-field maximum at distance of about λ/4 from the radiotelephone antenna's base, as shown in Fig. 3.7. The experiment was repeated with walkie-talkies working within 150 and 450 MHz bands.

Let's focus our attention on the fact that the maximal current intensity at the antenna base (and thus the operator's hand or the lip current) depends mainly on the device's output power, as the antenna's input impedance is usually standardized. The current may be estimated as follows:

Figure 3.7   A standing wave on the arm of a radiotelephone operator.

$$I_{max} = \sqrt{\frac{P_{fed}}{R_{in}}}$$

(3.14)

where:

P_{fed} = power fed to the antenna,

R_{in} = input impedance of the antenna.

After substituting into Eq. (3.14) typical values for the devices used in the experiment, i.e., P_{fed} = 5 W and R_{in} = 50 Ω, we will have the maximal magnitude of the current intensity flowing through the hand of the operator:

$$I_{max} = 0.3\ A$$

The estimated current intensity will appear at relatively low frequencies, where electrical sizes of the radiotelephone casing are much less than the wavelength. The measured currents are well below 0.3 A, which may be a result of the substantial role played by the casing, the measurement conditions, and the very limited accuracy of the measurement.

As may be seen from Table 3.2, the measured operator's hand and lip currents sometimes remarkably exceed magnitudes given in Tables 1.5 and 1.7 as permissible for a foot (leg)! It is not appropriate here to discuss the correctness (or incorrectness) of accepted legal regulations. However, the presented effect reveals the necessity to take into account not only theoretically estimated EMF energy absorption from a radio-telephone antenna (as it has been done until now) but also to take into consideration the conduction currents, I_h and I_l, while studying energy absorption in an operator's body and investigating the hazard created by these devices.

As far as current issues go, it is necessary to mention one more problem that is usually forgotten while estimating the exposure of a portable terminal operator. The problem is in VLF H-fields generated, consistent with the Biot-Savart law, by pulsed DC current supplying the power amplifier of the device, as shown in Fig. 3.8.

In many device designs, the power source connecting cables are quite long so as to connect battery compartment, usually located in the bottom of the device, with PA, as a rule mounted in its top. The magnitude of current (I) may be estimated using following formula [10]:

$$I = \frac{P}{\eta V}$$

(3.15)

where:

      P = declared HF power of the device,

      V = voltage,

      η = efficiency of the PA.

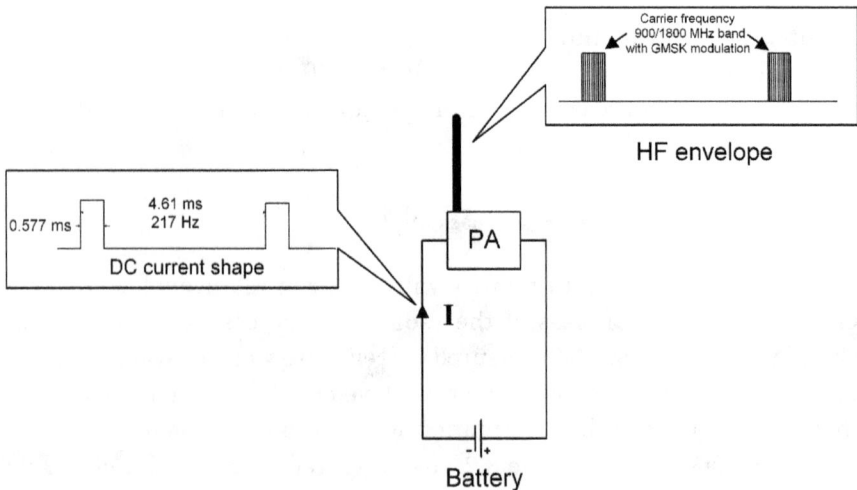

Figure 3.8 A power amplifier (PA) DC pulsed current.

Estimates on the basis of the Biot-Savart law, using Eq. (3.15), H-field strengths at a distance r = 1 cm from the surface of different types of handheld devices, for η = 0.7, are shown in Table 3.3.

Table 3.3 H-Field Close to Handheld Devices

Standard	P_{max} [W]	V [V]	I [A]	H [A/m]
GSM	2	6	0.48	7.65
	2	3.6	0.79	12.66
	2	2.4	1.2	19
TETRA	10	12	1.2	19
	10	7.2	1.98	31.6
	3	3.6	1.2	19
DECT	0.25	3	0.12	1.9

Results derived from measurements of different GSM and DECT type hand-held phones confirm correctness of these estimations. As an example, Figs. 3.9 and 3.10 present results of H-field measurements of a GSM terminal in both time and frequency domains.

Although biomedical interactions are not the subject of this work, it is worth mentioning two important problems related to mobile communication:

- It is not clear whether the use of a mobile radiotelephone (cellular phone) should be considered as a professional or nonprofessional exposure.

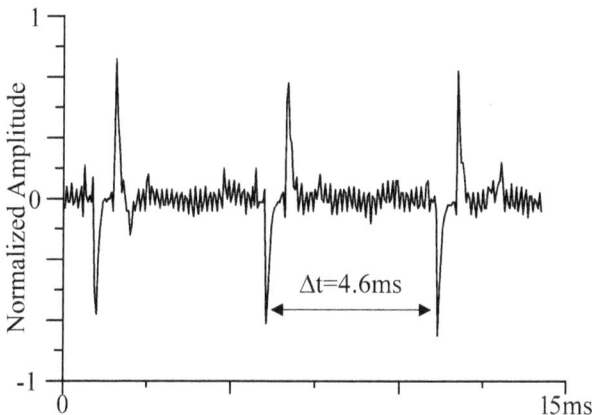

Figure 3.9 LF H-field near a GSM phone.

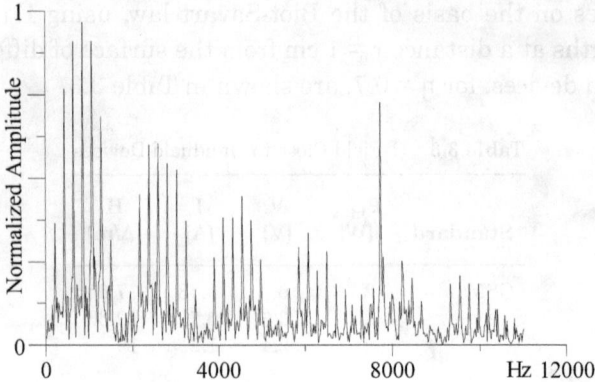

Figure 3.10   Spectrum of LF H-field of a GSM phone.

- The majority of the theoretical studies of EM energy absorption in an operator's body from hand-held radiotelephones is devoted to calculation of the absorbed energy and its distribution in the body (head), whereas the possibility of the EMF envelope detection suggests the need to take into account the *in vivo* role of detected VLF currents within the exposed body and direct induced currents by LF H-fields. Moreover, there are fields and currents over a wide frequency range generated by processing part of a device [11]. It is possible that neglecting all these currents results in differences between laboratory and epidemiological studies.

References

[1] D. A. Tschernomordik. *Estimation of the far field boundary of a symmetrical dipole* (in Russian). Trudy NIIR, No. 4/1972 pp. 55–60.

[2] C. C. Johnson, T. C. Rozzell. Liquid crystal fiber optic RF probes, part I. *Microwave Journal,* 1975, No. 8, pp. 55–57.

[3] C. C. Johnson, A. W. Guy. Nonionizing electromagnetic wave effects in biological materials and systems. *Proc. IEEE* 1970, Vol. 60, pp. 692–718.

[4] E. L. Larsen, R. A. Moore, J. Acevado.   A microwave decoupled brain-temperature transducer. *IEEE Trans.* Vol. MTT-22, 1974, pp. 438–444.

[5] C. A. Cain, M. M. Chen, K. L. Lam, J. Mullin. *The viscometric thermometer.* U.S. Dept. of Health, Education and Welfare HEW Publication (FDA) 78–8055, pp. 295–305.

[6] J. Y. Chen, O. P. Gandhi, D. Wu. Electric field and current density distributions induced in a millimeter-resolution human model for

EMFs of power lines. XVI-th Annual Meeting of the BEMS, Copenhagen 1994.

[7] A. W. Guy. Analyses of electromagnetic fields induced in biological tissues by thermographic studies of equivalent phantom models. *IEEE Trans.*, Vol. MTT-19, 1971, pp. 205–214.

[8] H. Trzaska. ARS and EM-radiation hazard. *Proc.1994 Intl. EMC Symp.* Sendai, pp. 191–194.

[9] P. Bienkowski, H Trzaska: Operator's body role in mobile radiotelephony. 12th Intl. scientific conference, Bratislava 2002, pp. 226–229.

[10] P. Bienkowski. The new GSM-like exposure system. *Proc. Intl. EMC Wroclaw Symp.* Wrocaw 2004, pp. 74–75.

[11] P.Bienkowski, H.Trzaska. Forgotten currents? *Proc. 6th Intl. Workshop on Biological Effects of Electromagnetic Fields.* Bodrum, Oct. 2010 (electronic version).

Chapter 4

Electric Field Measurement

The basic method of electric field measurement, over a wide frequency range, involves the use of charge induction phenomenon in a body illuminated by the field. As shown in Fig. 4.1, the electromotive force (emf) e_E induced by the electric component of the EMF generated by an arbitrary source in a symmetrical dipole antenna of total length 2h is:

$$e_E = \int_{-h}^{h} \mathbf{E} \; \mathbf{dh}$$

(4.1)

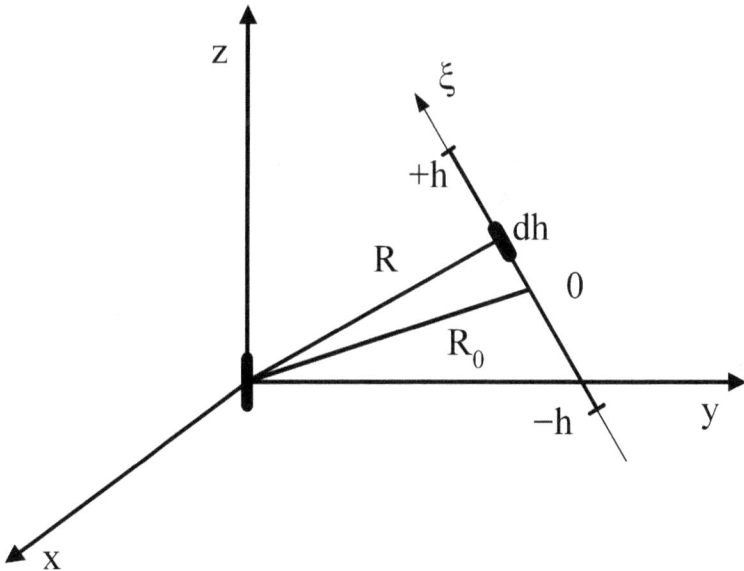

Figure 4.1 Symmetrical dipole in coordinate system.

If the source of the field is the elemental electric dipole placed in the coordinate system as shown in Fig. 2.4, and the length of the dipole the condition $1 \ll R_0$, then making use of Eq. (2.25), we may write:

$$e_E = C_1(\alpha) \int_{-h}^{h} \frac{\exp(-jkR)}{R^\alpha} \cos(\mathbf{E}, \mathbf{dh}) \, dh \tag{4.2}$$

where:

> R = the distance between the point of integration and the center of the coordinate system:

$$R = \sqrt{R_0^2 + \xi^2 + 2R_0 \, \xi \, \cos(\mathbf{E}, \mathbf{dh})} \tag{4.3}$$

where:

> ξ = current length of the antenna, $-h \leq \xi \leq h$,

$\cos(\mathbf{E}, \mathbf{dh}) = \mathbf{1}_E \times \mathbf{1}_h,$

$\mathbf{1}_E$ and $\mathbf{1}_h$ = versors of vector $\mathbf{E}$ and $\mathbf{h}$, respectively.

Equation (4.2) allows an estimation of the magnitude of the emf induced in the measuring antenna while the latter is arbitrarily located relative to the field source. With further consideration, the formula will be applied for E-field measurement error estimations, specifically for near-field measurements and for antenna (E-field probe) directional pattern synthesis.

4.1 FIELD AVERAGING BY A MEASURING ANTENNA

The maximal changes of amplitude and phase of the measured field along an antenna will appear for the radial component of the field, i.e., for $\mathbf{E}$ parallel to $\mathbf{h}$, then:

$$e_E = C_1(\alpha) \int_{R_0-h}^{R_0+h} \frac{\exp(-jkR)}{R^\alpha} \, dR \tag{4.4}$$

Making use of the Schwartz inequality, after integration, we have:

$$e_E \leq C_1(\alpha) \sqrt{\frac{\exp(-j2kR_0)\sin 2kh}{k}\left[\frac{1}{1-2\alpha}\left(\frac{1}{(R_0+h)^{2\alpha-1}} - \frac{1}{(R_0-h)^{2\alpha-1}}\right)\right]}$$

(4.5)

If the measuring antenna fulfills the conditions $h \ll R$ and $h \ll \lambda$, the emf induced in the antenna e'_E is given in Eq. (4.6). These conditions must be met in order to undertake calibration of the system.

$$e'_E = C_1(\alpha) \frac{\exp(-jkR_0)}{R_0} 2h$$

(4.6)

Dividing Eq. (4.6) by Eq. (4.5) gives:

$$\frac{e'_E}{e_E} \leq \sqrt{\frac{2kh}{\sin 2kh}\left[\frac{(1-2\alpha)2h}{R_0^2}\frac{(R_0^2-h^2)^{2\alpha-1}}{(R_0-h)^{2\alpha-1} - (R_0+h)^{2\alpha-1}}\right]}$$

(4.7)

or:

$$\frac{e'_E}{e_E} \leq \frac{kh}{\sin 2kh} + \frac{(1-2\alpha)h}{R_0^{2\alpha}}\frac{(R_0^2-h^2)^{2\alpha-1}}{(R_0-h)^{2\alpha-1} - (R_0+h)^{2\alpha-1}}$$

(4.8)

The error in the E-field measurement resulting from averaging the field measured by the measuring antenna is defined as:

$$\delta_E = \frac{e_E - e'_E}{e_E} \leq \delta_{1E} + \delta_{2E}$$

(4.9)

where:

$\delta_{1E} = $ an error expressing the field phase changes along the antenna:

$$\delta_{1E} = \frac{1}{2}\left(1 - \frac{2kh}{\sin 2kh}\right)$$

(4.10)

and δ'_{2E} = an error representing the amplitude changes along the antenna:

$$\delta'_{2E} = \frac{1}{2}\left[1 - \frac{(1 - 2\alpha)2h}{R_0^{2\alpha}} \frac{(R_0^2 - h^2)}{(R_0 - h)^{2\alpha - 1} - (R_0 + h)^{2\alpha - 1}} \right] \tag{4.11}$$

Equation (4.9) reflects measurement errors caused by two factors limiting accuracy: averaging the phase of the measured field along the antenna and averaging the amplitude. We will briefly analyze the first of them here, whereas the latter will be discussed in more detail in Section 4.3.

Field spatial distribution and field strength measurement requires the use of probes equipped with antennas whose sizes are much less than the wavelength of interest. This is especially true under conditions of multipath propagation and interference, and the measurement of the maximal and minimal magnitudes of the field strength (while standing waves appear). It would be most convenient to use "point antennas" (zero-dimensional ones), but since they are unavailable, we will consider the results of formal, though slightly overestimated, calculations of measurement errors due to the finite electrical length of the E-field probe's antenna. Equation (4.10) gives the error resulting from the phase changes along the antenna, dependent on kh. The formula makes it possible to find the magnitude of kh for which the value of the error will not exceed a permissible level. The error δ_{1E} is plotted in Fig. 4.2 as a function of kh.

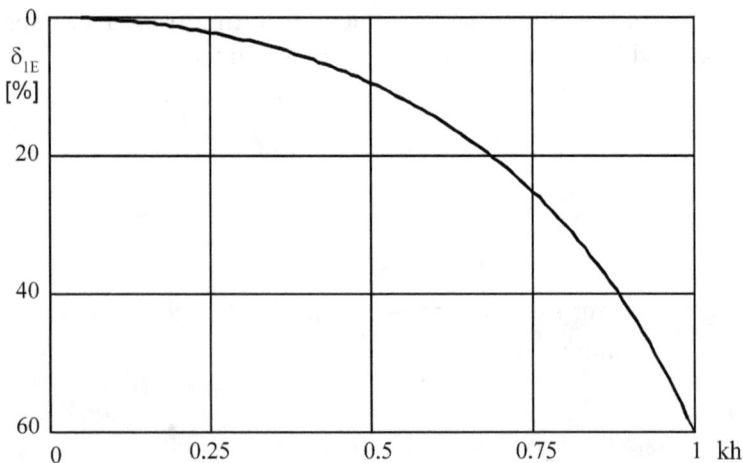

Figure 4.2   Error δ_{1E} versus kh.

It should be noted that the estimated magnitude of the error shown in Fig. 4.2 is about twice that defined by the ratio of the current intensity at the input of the antenna to that averaged along the antenna. In light of the above discussion, the use of antennas longer than h = λ/4 seems to be unacceptable. In the radio frequency ranges and lower, even a quarter wave antenna can be physically large and considerably exceed the size of the space where the measurement is to be performed. We may conclude that the electrical sizes of the antennas will become important when we are concerned with EMF measurements at decimeter waves and shorter.

4.2 INFLUENCE OF FIELDS FROM BEYOND A PROBE MEASURING BAND

As mentioned in Chapter 3, it is most advantageous to use wideband meters. However, because of both the accuracy of the monochromatic field measurement and the measurement of the field whose spectrum is unknown, it is important to know, as precisely as possible, the frequency response of the probe.

The typical wideband probe used for electric field measurement consists of a symmetrical dipole antenna loaded by a detection network. The detection network may be a diode, a thermocouple or an optical modulator whose output voltage is fed to an indicating device. As has been shown [1], without regard to the type of the detector applied, the probe may have certain maxima in its frequency response, especially at the highest frequencies. These result from the increase of the effective length of the antenna, resonances of the probe's parasitic reactances, variations of the detector efficiency with frequency, and other factors. However, the frequency response of the probe should be maximally flat within the entire measuring band, and it is assumed that outside the band, the sensitivity of the probe should not exceed that within the band. The latter suggests the use of RC low-pass filters that would artificially cut off the response at the highest frequencies on one side and permit its shaping, if desired, on the other. Figure 4.3 shows a simplified schematic diagram of the probe described. The probe is additionally equipped with a low-pass filter as mentioned, which is placed between the antenna and the detection diode; this allows shaping of the probe's response. Figure 4.4 presents its equivalent network for high frequencies.

If one assumes that, for dipole antennas with arms having a length h < λ/6, the input capacitance of the antenna C_A and its effective length

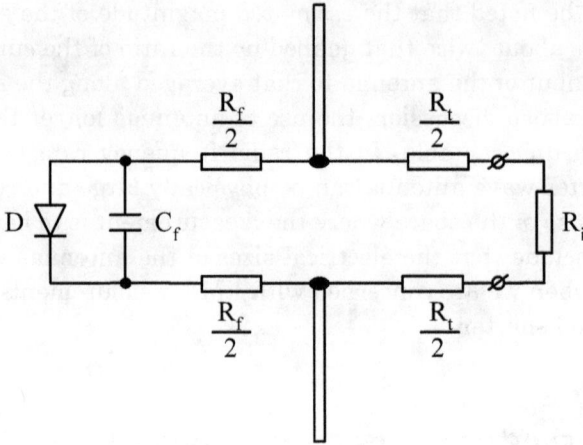

Figure 4.3 Simplified schematic diagram of the E-field probe.

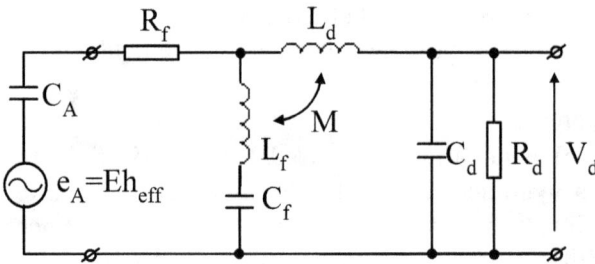

Figure 4.4 Equivalent circuit of the E-field probe for high frequencies.

h_{eff} are frequency independent, then the transmittance $T(jf)$ of the probe, defined as the ratio of the voltage delivered to the detector V_d to the emf e_E induced by the field in the antenna, is given by:

$$T(jf) = \frac{V_d}{e_E} = \frac{1 - \left(\dfrac{f}{f_3}\right)^2}{\dfrac{C_f}{C_A} + 1 - \left(\dfrac{f}{f_1}\right)^2 - \left(\dfrac{f}{f_2}\right)^2 + \left(\dfrac{R_f}{R_d} + \dfrac{C_d}{C_A}\right)\left[1 - \left(\dfrac{f}{f_4}\right)^2\right] + \left(\dfrac{f}{f_1}\right)^2\left(\dfrac{f}{f_2}\right)^2} +$$

$$+ j\left\{\left(\dfrac{f}{f_6} - \dfrac{f}{f_3}\right)\left[1 - \left(\dfrac{f}{f_3}\right)^2 + \dfrac{f}{f_7}\right] + Q_d\left[1 - (1 - K^2)\left(\dfrac{f}{f_3}\right)^2\right]\right\}$$

(4.12)

where:

$$f_i = \frac{\omega_i}{2\pi} \qquad i = 1 \div 7$$

$$\omega_1^2 = \frac{1}{L_d C_d}$$

$$\omega_2^2 = \frac{1}{L_f C_f}$$

$$\omega_3^2 = \frac{1}{C_f (L_f + M)}$$

$$\omega_4^2 = \frac{1}{C_f (L_f + L_d + 2M)}$$

$$\omega_5 = \frac{1}{C_A R_d}$$

$$\omega_6 = \frac{1}{C_d R_f}$$

$$\omega_7 = \frac{1}{C_f R_f}$$

$$Q_d = \frac{\omega L_d}{R_d}$$

$$K = \frac{M}{\sqrt{L_f L_d}}$$

$R_d =$ parallel connection of the diode resistance ρ and $R_t + R_i$.

Other indications are the same as in Fig. 4.3; in particular:

R_f, C_f = components of the RC low-pass filter,

R_t = resistance of the transparent line,

R_i = input resistance of a DC voltmeter,

C_d = equivalent capacitance of the detector.

Equation (4.12) allows the analysis of the probe transmittance in different configurations. The formula may be simplified if we assume M = 0, which is usually valid. The parasitic inductances may be omitted, and has very little effect on the transmittance, especially when it is limited at the higher frequencies. The resistance of the transparent line as well as that of the voltmeter sometimes may be neglected. Of course, they physically do not exist, for instance, when the antenna is loaded by an optical modulator. It makes it possible to apply the formula in its simplified form for the transmittance estimations of different versions of the E-field probe.

The transmittance given by Eq. (4.12) covers a wide frequency range that may be divided into three frequency bands:

1. The low-frequency band, where:

$$\left| T(jf) \right| = \frac{f}{f_5} \tag{4.13}$$

2. The medium frequency band (measuring band) with constant transmittance:

$$T(jf) = \text{const} = \frac{C_A}{C_A + C_d + C_f} \tag{4.14}$$

The corner frequencies of the band, at which the transmittance decreases by 3 dB, are defined as follows:

• The lower corner frequency: f_1, for $f \to 0$, while R_d is equal to the equivalent reactance of the probe:

$$f_1 = \frac{1}{2\pi R_d (C_A + C_d + C_f)} \tag{4.15}$$

or if we assume the transmittance's decrease in any degree δ:

$$f_1(\delta) = \frac{1 - \delta}{2\pi R_d (C_A + C_d + C_f) \sqrt{1 - (1 - \delta)^2}}$$

(4.15a)

· The upper corner frequency: f_u, for $f \to \infty$:

$$f_u = \frac{C_A + C_d + C_f}{2\pi R_f C_A (C_d + C_f)}$$

(4.16)

or, per analogy to Eq. (4.15a), for the transmittance reduction in δ:

$$f_u(\delta) = \frac{\sqrt{(1 - \delta)^2 - 1} \left(C_A + C_d + C_f \right)}{2\pi (1 - \delta) R_f C_A (C_d + C_f)}$$

(4.16a)

Equation (4.16) is true if $R_f \ll R_d$ and $f \ll f_{1-4}$,

3. The high frequency band, where $f \gg f_{1-4}$:

$$\left| T(jf) \right| = \frac{1}{f} \frac{f_6 f_7}{f_6 + f_7}$$

(4.17)

Equation (4.17) reflects the issue in highly simplified form, because it does not take into consideration the changes of the antenna's electric parameters, which may take place within the frequency band. However, it reveals the possibility of achieving the decreasing run of the transmittance modulus in a chosen frequency range if a low-pass filter (a detector of shaped frequency response) is used.

Even if we take into consideration the simplifications, applied when Eqs. (4.15) and (4.16) were introduced, they remain of primary practical importance. The formulas make it possible to select the detector and filter that allow us to obtain the flat character of the probe's frequency response within the desired frequency band using a single low-pass filter. In particular, they enable the selection of corner frequencies and the transmittance limitation in the upper band. The use of multiple filters enables us to achieve agreement between the probe's frequency response and the thresholds of a protection standard. However, it has to be said that the use of any filter reduces the probe's sensitivity.

Now we will outline more general considerations. As shown above, in our measurements we can apply only antennas that are much smaller as compared to the shortest wavelength of the measuring band. In terms of (1) the possibility of precisely calculating the probe's frequency response over the entire frequency range and, in particular, at frequencies above the measuring band, and (2) the possibility of using the probe at frequencies even exceeding those accepted so far (for instance, its use in the far field as well), we will also analyze the transmittance of the probe for h > λ/6. Additionally, we will consider the advantages that may result from using an n-segment RC filter. The transmittance of such a probe, whose equivalent network is shown in Fig. 4.5, is given by Eq. (4.18).

$$T = \frac{V_d}{E} = \frac{h_{eff}}{A_{11}}$$

(4.18)

where:

A_{11} = term of a matrix A given by:

$$[A] = [A_A] \cdot [A_{f1}] \cdot [A_{f2}] \cdots [A_{fn}] \cdot [A_d]$$

(4.19)

and:

$$[A_{fi}] = \begin{bmatrix} 1 + \dfrac{j\omega R_{fi} C_{fi}}{1 - \omega^2 L_{fi} C_{fi}} & R_{fi} \\ \dfrac{j\omega C_{fi}}{1 - \omega^2 L_{fi} C_{fi}} & 1 \end{bmatrix}$$

(4.20)

Index "i" denotes i-th segment of the filter, $1 \le i \le n$.

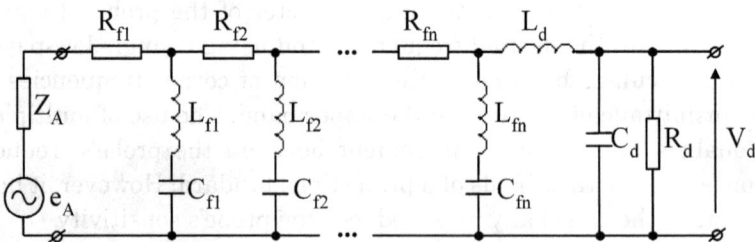

Figure 4.5 Equivalent network of the probe with an n-segment RC filter.

$$[A_d] = \begin{bmatrix} 1 + \dfrac{j\omega L_d}{R_d}(1 + j\omega R_d C_d) & j\omega L_d \\ \dfrac{1 + j\omega R_d C_d}{R_d} & 1 \end{bmatrix}$$

(4.21)

$$[A_A] = \begin{bmatrix} 1 & Z_{11} \\ 0 & 1 \end{bmatrix}$$

(4.22)

where: Z_{11} = input impedance of an antenna given by Eq. (4.33).

The effective length of a thin symmetrical dipole antenna with a quite large ratio of length to diameter is [2]:

$$h_{eff} = \frac{2 \left[J_0(kh) - \cos kh \right]}{k \sin kh}$$

(4.23)

where: J_0 = Bessel function of the first kind and zeroth order.

Examples of calculated frequency responses of the E-field probe are shown in Fig. 4.6. The curves are normalized in the relation to that for n = 3.

The upper diagram shows results of calculations performed with the assumption that $L_f = 0$, whereas the lower diagram is for $L_f \neq 0$. Curve 1 shows the transmittance of the probe without a filter, curves 2 and 5 represent the probe with a single-segment filter, curves 3 and 6 refer to a double filter and curves 4 and 7 to a triple one. The calculations were done for similar time constants of the filters.

From the curves shown in Fig. 4.6, we can draw the conclusion that the use of multi-segment filters makes it possible to reach the desirable frequency response shape and large attenuation within the high-frequency range. The conclusion is evident in some sense, but its use is strongly limited. This is primarily because of a decrease of the probe's sensitivity in proportion to the number of filter segments applied in the probe. The transmittance of the probe with an n-segment filter, within the medium frequency band, is given by:

$$T(jf) = \frac{C_A}{C_A + C_d + \displaystyle\sum_{i=0}^{n} C_{fi}}$$

(4.24)

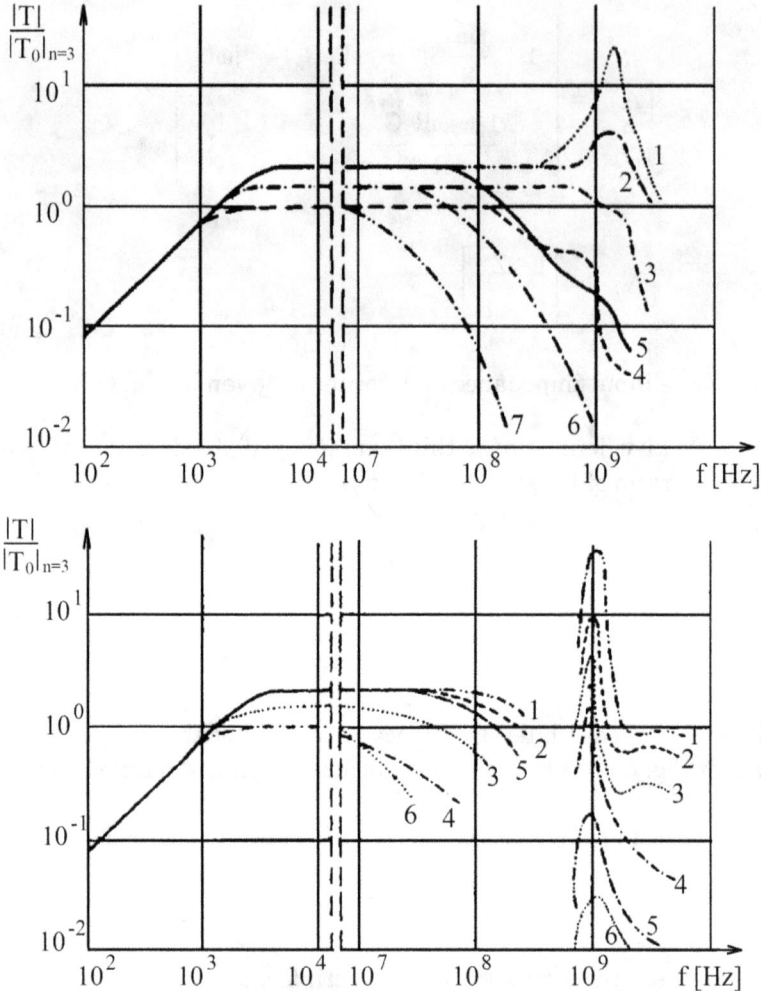

Figure 4.6 Calculated transmittances of the E-field probe with n-segment filters.

Thus, the larger sum of the filters' capacitances, the smaller the transmittance and, consequently, the meter's sensitivity. Even a single-segment filter in the probe permits us to obtain its transmittance shape at the highest frequencies such that the sensitivity, above an arbitrarily selected upper corner frequency, won't exceed that within the measuring band. Simultaneously, we must remind ourselves that using even a single-segment filter causes a reduction in the probe's sensitivity, which results directly from Eqs. (4.14) and (4.24).

The results of the above calculations represent one case of the cascade connection of several identical filters. The use of multi-segment filters of different resonant frequencies permits us to design the frequency run of the transmittance similar to that of the frequency-dependent thresholds given by a protection standard. The filters also allow construction of a probe with several bands in which the transmittance will be frequency independent and of different (desirable) magnitudes.

By using these filters as well as traps (band-stop or notch filters), tuned to the antenna's resonant frequencies, it is possible to design a super-wideband E-field probe that will have flat frequency response at frequencies corresponding to the length of the applied antenna, limited by h < 0.5λ. It would be possible to successfully continue this approach to obtain an acceptably flat response; however, above the limit, the directional pattern of the antenna splits, and the number and magnitude of the lobes in the pattern increases. This excludes the use of such an antenna (with no regard, of course, to the previous discussion of the antenna's size limitation). Even if the shape of the directional pattern in some cases may be permissible, the synthesis of an E-field probe with a spherical directional pattern (omnidirectional pattern) and acceptable pattern irregularities seems impossible.

A schematic diagram and its equivalent circuit of a probe with a trap are shown in Fig. 4.7, whereas the calculated frequency responses are shown in Fig. 4.8. Curve 1 shows the transmittance while the trap was not tuned to the antenna's resonant frequency and the quality factor of the trap was too high. In curve 2, the trap was tuned but still with unchanged quality factor. Curve 3 represents optimal compensation of the resonant effect. The probe may be used principally when super-wideband measurements are performed in the far field and with the use of a broadband receiver or a spectrum analyzer, when the detection diode is replaced by a light modulator.

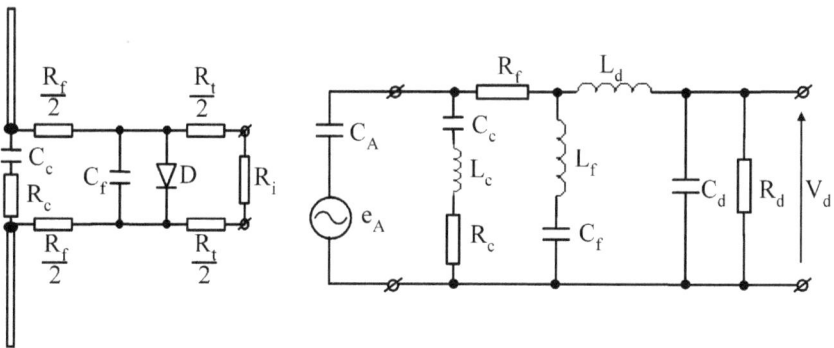

Figure 4.7 Schematic diagram and equivalent network of the E-field probe with an RC filter and a trap $R_c C_c$.

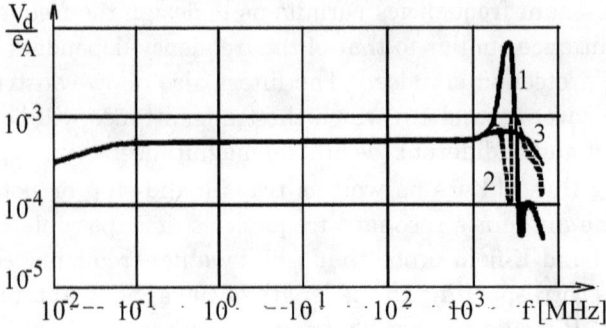

Figure 4.8 Normalized transmittance of the probe with an RC filter and a trap.

In the above examples, the detector was represented by linear elements. This made it possible to simplify the analyses, but at the expense of a certain decrease in precision of the estimations. This restriction mostly reflects the diode detector, a bit less than with the thermocouple, and it does not concern the optical modulator. Currently, however, the diode detectors are the most popular choices in this field. Because of the dependence of the nonlinear element parameters on its chosen working point, especially the lower corner frequency, it may be a function of the intensity of the measured field [3]. However, the main aim here was to show the possibilities and necessities of the probe's frequency response shaping and possible errors that could result when performing measurements with probes of uncontrolled response, even if only a single source is active during the measurements. The errors may be caused by the harmonic frequencies' presence in the spectrum radiated by the source or the power-line frequency field and radiation caused by other unexpected sources (for instance, ultrasonic generators, HF power sources exciting high-power lasers, information equipment, and nearby broadcast and TV stations). Such unwanted sources can interfere with the measured one and, consequently, may affect the measurement results.

Now let's consider a case wherein a number, n, of segments of low-pass filters is ranging to infinity (n → ∞) and, moreover, the filters create a resistive antenna. This concept was proposed by authors from the National Institute of Standards and Technology (NIST), and several designs for different applications were proposed [5,6].

All of the NIST probe models were designated for wideband EMF measurements. An 8-mm long, resistively tapered dipole (Fig. 4.9) with a selected low-barrier Schottky diode detector, working within a frequency range of 100 kHz to 18 GHz, was then used as a transfer stan-

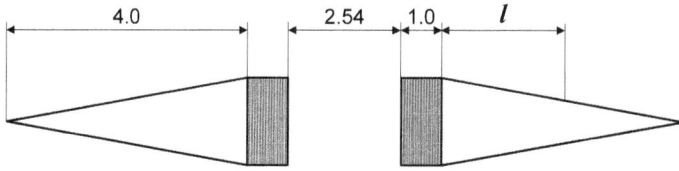

Figure 4.9 NIST resistive dipole (dimensions in millimeters).

dard in international comparison of EMF standards, organized and headed by the NIST. In the model, a resistive foam, made of tantalum nitride, was sputtered on a dielectric substrate. A profile of the resistive substrate, as a function of distance to the center of the dipole (1), is shown in Fig. 4.10.

Without regard to the application, the probe's design makes it possible to be used for any measurements where a small EMF probe is necessary. The concept is utilized in some meters that are available on the market. The main advantages of the construction are the ability to increase measurement sensitivity and a flat (within ±2 dB) frequency response over the entire frequency range. Regardless of this, the problems related to mutual interactions with surroundings, field integration, and problems with the spherical radiation pattern are similar to other probes.

So far, we have discussed problems related to the construction of a probe with a wide measuring frequency band, effects that may appear at the highest frequencies, and the possibility of shaping the frequency response of a probe in its upper frequency range. Next we should focus

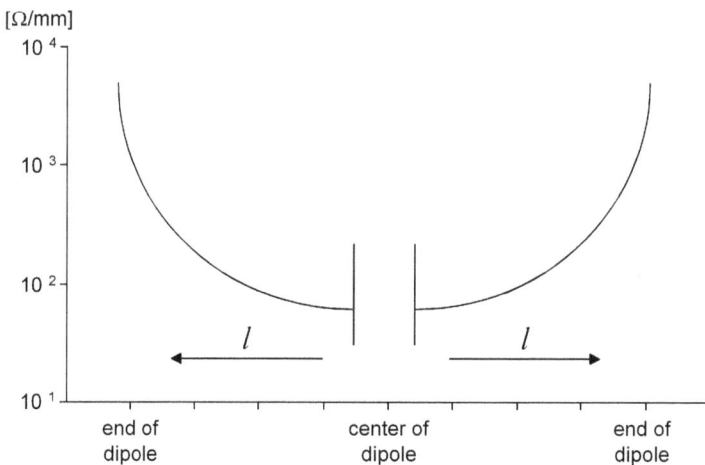

Figure 4.10 The resistive tapering profile of the dipole.

our attention on problems and effects that may appear at the lowest frequencies.

It is interesting to note that some meter manufacturers suggest that their meters can be used at frequencies far below their lower corner frequency. They provide a correction factor, taken from the frequency response as spelled out in the manual. We should note that almost all of these meters are equipped with probes that include nonlinear detectors (diode or thermocouple one). Usually, while no filters are in the probe or a single filter is used, the slope of the transmittance in the low-frequency range is 6 dB/oct. The slope is valid for a linear detector (which was assumed in the above calculations of the frequency response) whereas, for a square-law detector, the slope will be twice as much. As an example, Fig. 4.11 presents two frequency response graphs of the same probe measured with different applied field intensities [7].

While the detector works in its linear or nonlinear range, it is impossible to determine whether an error occurs. Therefore, it is imperative to drop this approach. The gross error may be the result of complete impossibility of evaluating the importance of the source's harmonics in the meter's indication (the sensitivity, in the range, increases proportionally with the harmonic's order or its square), the mentioned dependence of the frequency response on the amplitude of the measured field and the detector's working point, the temperature dependence of the detector's parameters, and other issues. These problems will be discussed in greater detail in Chapter 8.

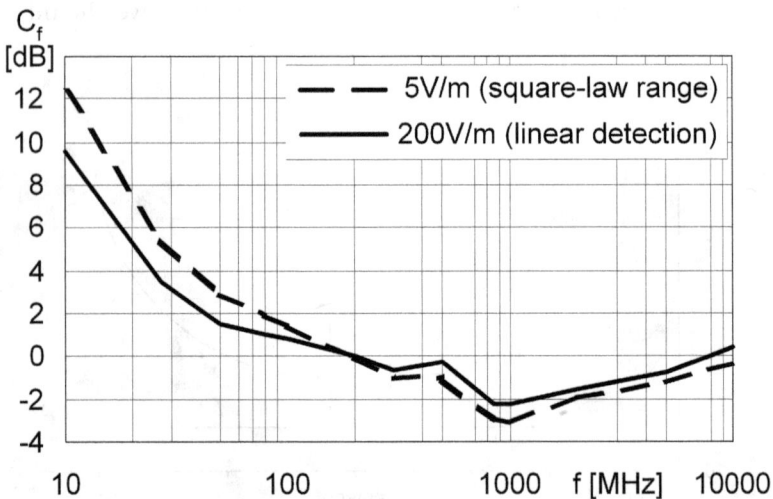

Figure 4.11 Frequency response of an EMF probe for square-law and linear detection. (C_f = calibration factor in relation to medium frequency range.)

The problem of the lower corner frequency has more parameters than presented above. It includes the following issues:

1. The lower corner frequency is given by Eq. (4.13). It may be deduced from the formula that the most important role in the limit is played by the equivalent resistance of the detector. Apart from the above-mentioned different frequency response at the lowest frequencies while the detector works in its linear and square-law range (Fig. 4.11), three phenomena may appear here:

   • The resistance is different, even in different specimen of detectors of the same type.
   • The resistance is a function of temperature.
   • The resistance may be affected by aging effects and overloading.

2. The transmittance decrease of an E-field probe in the low-frequency range (if we assume linear detection) is 6 dB/oct. Thus, attenuation of signals at the power line frequency decreases a bit (20 dB), and it may be insufficient, especially when probes for very low frequencies are in use. An example here may be a case of an induction heater working in kilohertz range. Its power amplifier is fed from a high-voltage source; moreover, its output power often may infiltrate the power line. Thus, an EMF measurement close to such a device, using VLF EMF probe, may be affected by the presence of a 50/60 Hz field (or even higher, if multiphase rectifiers are used in a power source).

A way to stabilize the frequency response against detector parameter variations, as well as decreasing a probe's attenuation at the power line frequency, is through the use of an additional high-pass filter in the probe as shown in Fig. 4.12 [8]. Apart from the previously described

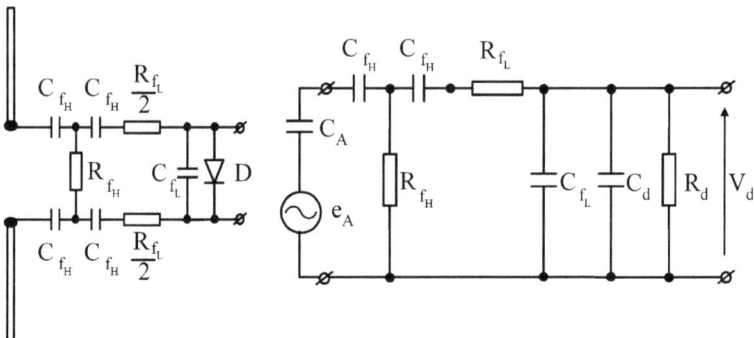

Figure 4.12 VLF E-field sensor with high-pass filter in antenna circuit.

low-pass filter (R_{fl}, C_{fl}), a high-pass filter (R_{fh}, C_{fh}) is added, and its resistance R_{fh} may dominate in the low-frequency limitation, as given by Eq. (4.15) and its definition of f_5.

Figure 4.13 presents the normalized results of frequency response measurements of the probe shown in Fig. 4.12 with and without a double high-pass filter.

Although the problems presented in this work are focused on the HF EMF measurements (where HF here may even mean several hertz or less), we would like to present an electrostatic E-field meter design proposed by the authors. An electrostatic field is not measurable using the probes and methods presented here. In order to measure the field, it is necessary to employ an artificial "alternation" (modulation) measured field. That is to develop a solution in which a probe is exposed to pulses of the field. In standard constructions, this is realized by a rotating screen, which periodically screens the probe, and in this way it is exposed to a pulsed field. The main disadvantage of this solution is the presence of the rotating element and the required precise stabilization of the rotation velocity—not to mention the power consumption, stability of the device, and its hybrid mechatronic construction. The proposed solution is shown in Fig. 4.14.

A probe is connected to an input amplifier while the input is periodically grounded by short pulses from a generator. This causes the input

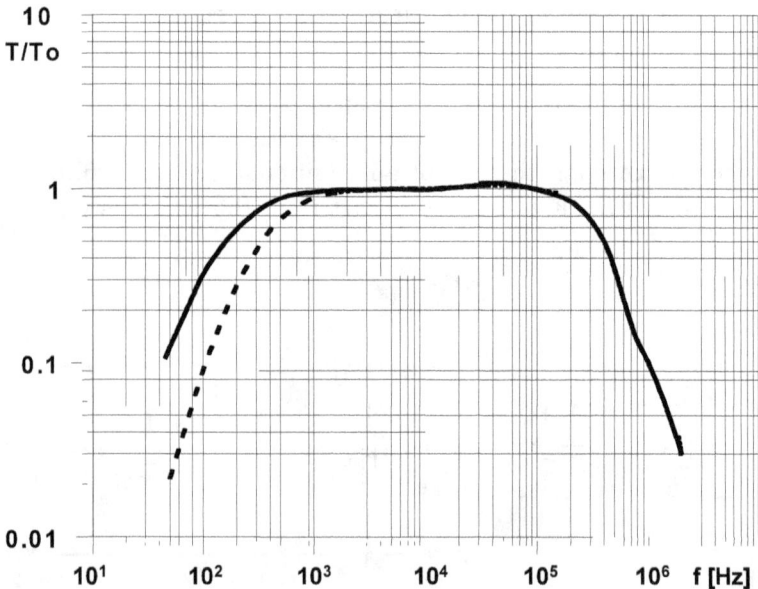

Figure 4.13 Frequency response of VLF probe without (solid) and with high-pass filter (dashed).

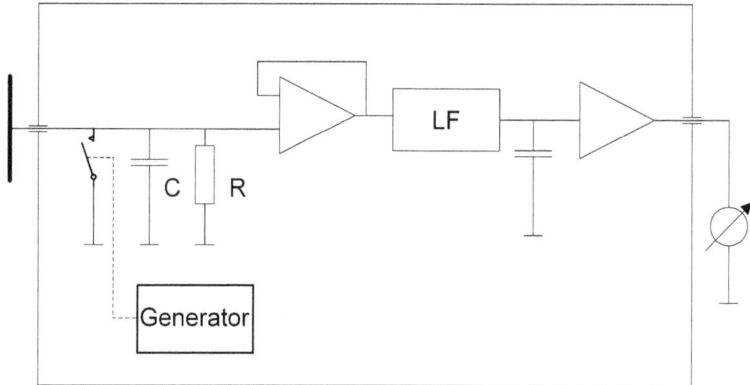

Figure 4.14 Block diagram of the proposed electrostatic field meter.

capacitor to be periodically charged and discharged. This construction would be sensitive to EMF from wide frequency range, and especially to fields of power line frequency. In order to limit the sensitivity of the device, the time-constant of an RC network is selected equal to the keying frequency. A front-end amplifier, which is used as a matching amplifier of unity gain amplification, feeds into a low-pass filter (LF). Then, after integration, the signal is fed to an indicator. The concept has been proven in several versions. The most sophisticated of these implementations contains two probes that allow symmetric measurement. However, the basic idea is identical, and it is based upon a keying of the device's input.

4.3 MUTUAL INTERACTION OF THE MEASURING ANTENNA AND THE FIELD SOURCE

4.3.1 Dependence of the Accuracy of E-Field Measurement on the Structure of the Source

Measurement accuracy limitations due to field averaging by the measuring antenna were discussed in Section 4.1. Those considerations allowed us to define the errors δ_{1E} and δ'_{2E}. The former was a function of the h/λ ratio and represented the influence of the phase variations along an antenna, while the latter was dependent on R_0/h and α. Thus, it represents, in terms of the accuracy of measurement estimations, both the relative distance between the antenna and the source and the structure of the source. As mentioned earlier, because of simplifications accepted when Eq. (4.11) was introduced, the calculated error when using the formula was overestimated by almost a factor of two. In order to make these estimations more precise, we will repeat the calculations, assuming (with an accuracy to δ_{1E}) that the phase variations of the measured field along the antenna do not take place; in other words, the term exp(–jkR) in Eqs. (4.4) and (4.5) is equal to exp(–jkR$_0$). This means that the measuring antenna is electrically short enough and that the emf induced in the antenna is averaged along the antenna:

$$e_{E} = C_{1}(\alpha) \, \exp\,(-jkR_{0}) \int_{R_{0}-h}^{R_{0}+h} \frac{dR}{R^{\alpha}}$$

(4.25)

and:

$$e'_{E} = C_{1}(\alpha) \, \exp\,(-jkR_{0}) \, \frac{2h}{R_{0}^{\alpha}}$$

(4.26)

Making use of Eqs. (4.25) and (4.26), we now define the error δ_{2E} as:

$$\delta_{2E} = \frac{e_{E} - e'_{E}}{e_{E}}$$

(4.27)

Substituting Eqs. (4.25) and (4.26) in to Eq. (4.27), we obtain:

• for $\alpha = 3$

$$\delta_{2E} = 1 - \frac{\left(R_{0}^{2} - h^{2}\right)^{2}}{R_{0}^{4}}$$

(4.28)

• for $\alpha = 2$

$$\delta_{2E} = \frac{h^{2}}{R_{0}^{2}}$$

(4.29)

- and for $\alpha = 1$

$$\delta_{2E} = 1 - \left[\frac{R_0}{2h} \ln \frac{R_0 + h}{R_0 - h} \right]^{-2}$$

(4.30)

Regardless of how we approach the calculations for the plane wave, for which $\alpha = 0$, we have $\delta_{2E} \equiv 0$.

The errors, calculated using Eqs. (4.28) through (4.30), are shown in Fig. 4.15.

The curves illustrate the influence of the structure of the source upon E-field measurement accuracy as a function of R_0/h. We must stress here that the effect of averaging the measured field provides rigorous limits to the size of the antenna selected for measurements in close proximity to a primary or secondary source. The resultant accuracy of the field measurement in the near field, in the majority of cases, is dominated by the error. The error is particularly characterized by its rapid decrease with increasing distance, as may be deduced directly from Fig. 4.15.

4.3.2 Dependence of Accuracy on Antenna Input Impedance Changes

The mutual impedance of a thin dipole antenna, made of a perfect conductor of diameter $2a$ and located parallel at distance $b/2$ from a flat, infinitely large and perfectly conducting plane and its mirror reflection is given by Eq. (4.31) [4].

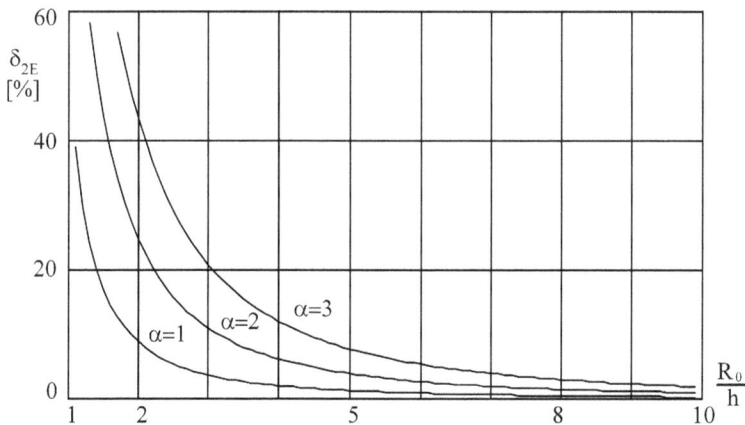

Figure 4.15 Error δ_{2E} as a function of R_0/h for $\alpha = 1$, 2, and 3.

$$Z_{21} = 30 \left[\left(2 + e^{-j2kh}\right) \left(\text{Ci } kx - j \text{ Si } kx\right) \Big|_{\sqrt{h^2 + b^2} - h}^{b} \right.$$

$$+ \left(2 + e^{j2kh}\right) \left(\text{Ci } kx - j \text{ Si } kx\right) \Big|_{\sqrt{h^2 + b^2} + h}^{b}$$

$$+ e^{j2kh} \left(\text{Ci } kx - j \text{ Si } kx\right) \Big|_{\sqrt{h^2 + b^2} + h}^{\sqrt{4h^2 + b^2} + 2h}$$

$$\left. + e^{-j2kh} \left(\text{Ci } kx - j \text{ Si } kx\right) \Big|_{\sqrt{h^2 + b^2} - h}^{\sqrt{4h^2 + b^2} - 2h} \right]$$

$$(4.31)$$

Taking into consideration the limit:

$$\lim_{b \to a} Z_{21} = Z_{11}$$

$$(4.32)$$

and taking into account that the problem is much smaller when compared to the wavelength of the considered field (i.e., for a, b, h << λ, neglecting the real part of the impedance as much smaller in comparison with the imaginary one), we have the input reactance of a short dipole in the free space X_{11}. After taking the limit of the impedance given by Eq. (4.31) for λ → 0 and omitting its real part, we obtain the mutual reactance of two dipoles placed at distance b apart. The demanded input reactance X_i of the dipole placed parallel to the perfectly conducting medium at distance b/2 we obtain as the difference of the two reactances, i.e.:

$$X_i = X_{11} - X_{21}$$

$$(4.33)$$

For free space (i.e., a space without any material objects that could affect the input capacitance of the antenna, $X_{21} \equiv 0$), the transmittance of the E-field probe, within the measuring band as given by Eq. (4.14), we now rewrite in the form:

$$T(jf) = \frac{X_d}{X_d + X_{11}}$$

$$(4.34)$$

where X_d denotes the equivalent reactance represented by the detection network—i.e., the input reactance of the detector, the sum of reactances of the filters applied, and the parasitic capacitances of the probe.

If the influence of the material media on the antenna's input reactance cannot be neglected (in other words, when $X_{21} \neq 0$), the transmittance is:

$$T'(jf) = \frac{X_d}{X_d + X_i}$$

(4.35)

Because the probe is calibrated under conditions where $X_{21} \approx 0$, when the measurements are performed in a neighborhood of a conducting medium (represented, in the estimations carried out, by the infinitely large and perfectly conducting plane), there will be a measurement error caused by the probe's antenna input impedance change, which is affected by the presence of the medium. We will define the error in the form:

$$\delta_{3E} = \frac{T'(jf) - T(jf)}{T'(jf)}$$

(4.36)

The input impedance of the antenna, as well as the mutual impedance of the antenna and its mirror image reflection, is a function of its slenderness ratio [2]. The slenderness depends on the ratio of the antenna length to the diameter of the conductor, i.e., h/a. Thus, it is necessary to introduce slenderness as a parameter in our considerations. Figure 4.16 shows the error δ_{3E} versus b/2h for h/a = 30, 300, and 1000, respectively.

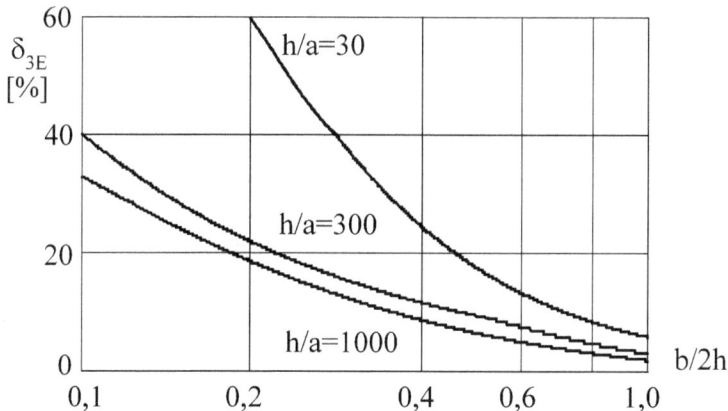

Figure 4.16 Error δ_{3E} versus b/2h.

Consider the three features that characterize the results:

- Errors δ_{2E} and δ_{3E} are functions of the probe–source distance, but their comparison shows that error δ_{3E} usually substantially exceeds δ_{2E}.

- The above comment becomes more critical if we take into account that the estimations were carried out for an antenna located parallel to an infinitely large conducting medium. However, when near the conducting medium (which in the calculations represents a primary or a secondary source of radiation), the dominant role is played by the radial E-field component, and the measuring antenna should be spatially oriented in accordance with the component. Thus, the presented estimations give results majorizing any error of this kind that could appear in the actual measurements.

- The latter is valid even if we take into account that estimations of δ_{2E} were obtained for an infinitely thin dipole, and the magnitude of the error will increase for dipoles of finite thickness.

To summarize these considerations, the estimated magnitude of errors as compared to actual errors that may appear during the measurement process is overstated. Thus, the estimation should be understood here as an illustration of possible worst-case inaccuracy of measurement caused by the separate limiting factors.

4.4 CHANGES IN THE PROBES' DIRECTIONAL PATTERN

It is important in near-field EMF measurements, where there are usually three spatial components of the field to be able to measure all three field components. This is as a result of arbitrarily oriented circular or elliptical polarization. This may occur, for instance, as a result of multipath propagation and interference from numerous primary and secondary sources. Of course, such measurements may be realized using a meter equipped with a probe with a single dipole antenna sequentially oriented to measure the individual spatial components and then finding the total field using simple calculations. However, the procedure is far from ideal and may lead to gross errors and mistakes.

By intuition, it may be supposed that design of a probe that would be sensitive to the three spatial components of the field should include an antenna system containing three mutually perpendicular dipole antennas, as shown in Fig. 4.17.

The emf induced in separate dipoles by an arbitrarily located E-field vector may be calculated using Eq. (4.1). For i-th dipole, it is:

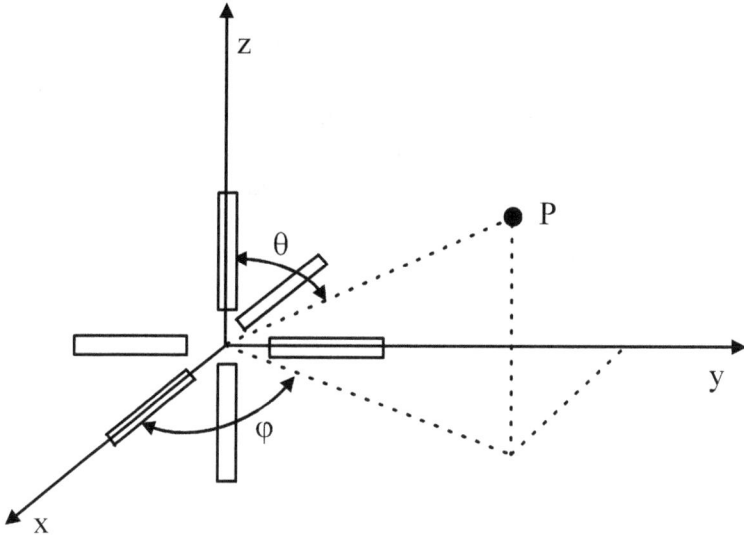

Figure 4.17 A set of three mutually perpendicular dipoles.

$$e_{Ei} = \int_{-hi}^{hi} \mathbf{E} \cdot \mathbf{dh}_i$$

(4.37)

If in Eq. (4.37) we replace "i" by x, y, and z, we will obtain the electromotive forces induced in dipoles that are located on the respective axes. If we then assume that the antennas are of identical size (i.e., $h_i = h_x = h_y = h_z$), that they are much shorter as compared to the wavelength, and that they are illuminated by a homogeneous monochromatic plane wave, with accuracy to a constant A, the sum of the voltages induced in these antennas may be expressed in the form:

$$e_E = \sum_{i=x}^{z} e_{Ei} = A \left[\cos (\mathbf{E,hx}) + \cos (\mathbf{E,hy}) + \cos (\mathbf{E,hz}) \right]$$

(4.38)

Because the formula should be valid for any spatial localization of the vector **E**, let's assume, for instance, that its direction coincides with that of the polar vector **R** of the spherical coordinate system (Fig. 4.1). Thus, the directional cosines of the vector **E** are identical to those of vector **R**, i.e., $\cos \theta$, $\sin \theta \cos \varphi$, and $\sin \theta \sin \varphi$.

When arbitrarily polarized field measurements are performed, it is desirable for the probe to be insensitive to the kind of the polarization and the direction of the **E** (or **H**) vector spatial rotations. To fulfil this condition, we would need the emf induced in the measuring antenna to be independent of the spatial coordinates of the **E** or **H** vector. For the antenna system shown in Fig. 4.17, we may intuitively suppose that these requirements are to be fulfilled, as separate antennas should be sensitive to specific spatial components of the **E** vector. To show that this is really true, it is enough to prove that the sum of directional cosines in the square brackets of Eq. (4.38) is a constant value. We will use a simpler method: we will equate Eq. (4.38), after its mentioned modification is completed, to zero. In a non-trivial case (A ≠ 0), if the hypothesis is true, the following equation must not be correct:

$$\cos \theta + \sin \theta \cos \varphi + \sin \theta \sin \varphi = 0$$

This formula represents an equation of a plane that includes the center of the coordinate system and intersects all the axes of the system at identical angles. Any change of the **E** vector localization (which is equivalent to the change of signs of separate terms in the equation) in relation to the coordinate system results only in the spatial rotation of the plane. Thus, there is a non-vanishing set of uniplanar **E** vectors on the plane that are "invisible" to the considered system of three mutually orthogonal antennas whose output voltages are directly added. It is easy to notice, however, that the separate terms in the formula added after squaring result in unity. That means that the sum is independent of the spatial coordinates of the considered vector, and the approach gives an answer to our search for the directional pattern of a probe that is insensitive to the spatial location of the **E** vector, i.e., an omnidirectional pattern or spherical directional pattern.

Chapter 7 is devoted to the considerations related to the synthesis of the probes with spherical pattern. Here we will only analyze the errors resulting from the pattern changes when measurements with the use of an omnidirectional probe are performed in the near field.

Standardized directional patterns of three identical dipoles, whose arms are of length h, located on separate axes of a Cartesian coordinate system (Figure 4.17) are given by:

$$f_x(\theta,\varphi) = \frac{\cos (kh \sin \theta \cos \varphi) - \cos kh}{\sqrt{1 - \sin^2 \theta \cos^2\varphi}} \tag{4.39}$$

$$f_y(\theta,\varphi) = \frac{\cos (kh \sin \theta \sin \varphi) - \cos kh}{\sqrt{1 - \sin^2 \theta \sin^2 \varphi}} \tag{4.40}$$

$$f_z(\theta,\varphi) = \frac{\cos(kh\cos\theta) - \cos kh}{\sin\theta}$$

(4.41)

If the parameters of the dipoles are identical and they are loaded by square-law detectors of identical efficiency, then, by summation of their output voltages, we will calculate the directional pattern of the system $F(\theta, \varphi)$:

$$F(\theta,\varphi) = f_x^2(\theta,\varphi) + f_y^2(\theta,\varphi) + f_z^2(\theta,\varphi)$$

(4.42)

We define the error of the measurement, resulting from irregularities of the system pattern:

$$\delta_{4E} = 1 - \frac{F_{min}}{F_{max}}$$

(4.43)

where:

F_{min}, F_{max} = respectively, the minimal and the maximal value of the function given by Eq. (4.42).

The magnitude of the irregularities of the directional pattern, calculated versus kh using Eq. (4.43), is shown in Fig. 4.18. This is of particular concern for antennas that are not electrically short. In our field, such antennas may be sometimes used, especially at microwave frequencies and in some applications related to the electromagnetic compatibility measurements. Based on Eq. (4.43) and Fig. 4.18, we can

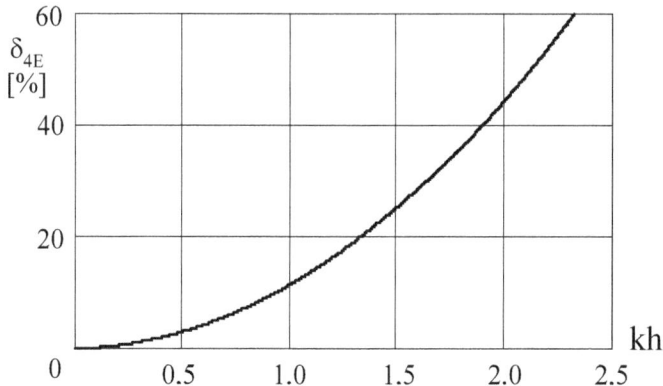

Figure 4.18 Error δ_{4E} versus kh.

estimate the maximal permissible electrical length (h/λ) of the antennas; for shorter antennas, the irregularities of the pattern should not
exceed those assumed.

Considerations relating to omnidirectional pattern irregularities are,
in some sense, similar to the above-presented estimations of measured
field averaging by a measuring antenna. As before, we may assume that,
for an antenna fulfilling the condition h << λ, the phase changes along
the antenna (here with an accuracy to δ_{4E}) are of negligible importance
with regard to the deformations of the directional pattern. Further considerations, also as before, will be limited to the influence of the field
amplitude changes along the antenna. Making use of Eq. (4.2), we will
describe the emf induced by a point source located at point P in three
mutually perpendicular dipoles whose directions coincide with those of
separate axes of the Cartesian coordinate system. Thus:

$$e_{Ex} = C_1(\alpha) \ \exp \ (-jkh) \int_{-h}^{h} \frac{\cos (\mathbf{E}, \mathbf{hx})}{R_x^{\alpha}} \, dx \tag{4.44}$$

$$e_{Ey} = C_1(\alpha) \ \exp \ (-jkh) \int_{-h}^{h} \frac{\cos (\mathbf{E}, \mathbf{hy})}{R_y^{\alpha}} \, dy \tag{4.45}$$

and:

$$e_{Ez} = C_1(\alpha) \ \exp \ (-jkh) \int_{-h}^{h} \frac{\cos (\mathbf{E}, \mathbf{hz})}{R_z^{\alpha}} \, dz \tag{4.46}$$

where:

e_{Ex}, e_{Ey}, e_{Ez} = emf induced by the field in dipoles located on axes x, y, z
that here represent non-standardized directional pat-
terns of the three short dipoles,

R_x, R_y, R_z = distances between point P (R_0, θ, φ) and dx, dy, and dz,
respectively:

$$R_x = \sqrt{R_0^2 + x^2 - 2R_0 x \sin \theta \cos \varphi}$$

$$R_y = \sqrt{R_0^2 + y^2 - 2R_0 y \sin \theta \sin \varphi}$$

$$R_z = \sqrt{R_0^2 + z^2 - 2R_0 z \cos \theta}$$

If we sum the squares of the emf induced in the three dipoles, we will obtain the emf $e(\theta, \varphi)$, which represents a non-standardized directional pattern of the system:

$$e\,(\theta,\varphi) = e^2_{Ex} + e^2_{Ey} + e^2_{Ez}$$

(4.47)

The synthesis of the spherical pattern of the system in the far field (or under calibration conditions) is determined by the nature of the antennas and their detectors. We should assume that the conditions can be fulfilled with the required accuracy. The deformation of the pattern, resulting from the specific conditions of the field measurement in the near field, in particular from the finite curvature of the field, we define (with accuracy to δ_{4E}) as:

$$\delta_{5E} = 1 - \frac{e_{min}}{e_{max}}$$

(4.48)

Calculated deformations of the pattern δ_{5E} as a function of R_0/h for $\alpha = 1$, 2, and 3 are plotted in Fig. 4.19.

The pattern is also affected by the input impedance changes of the separate antennas, which is a result of their proximity to a conducting medium. In order to estimate a role of the effect, it would be necessary to follow considerations presented earlier in Section 4.3.2 for an omnidi-

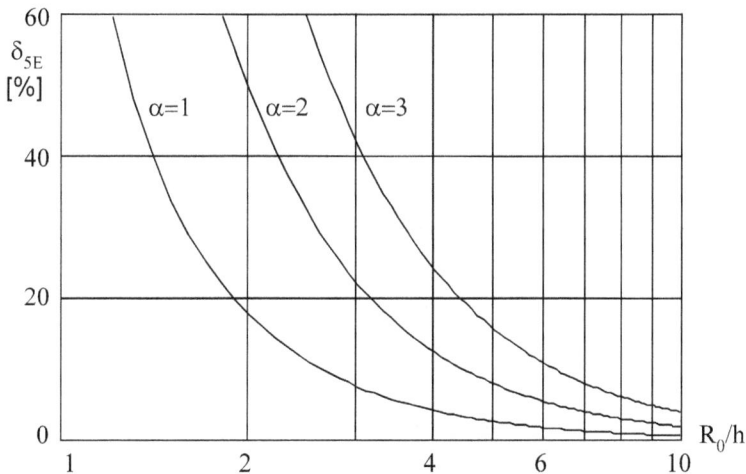

Figure 4.19 Error δ_{5E} versus R_0/h.

rectional probe. However, the section ended with a comparison of errors δ_{2E} and δ_{3E} and with the conclusion that the former usually remarkably exceeds the latter. Thus, we assume here that, in the case of the pattern deformations near material media, a similar conclusion is valid. We therefore will omit further discussion of the factor, but the reader should remember that this does not mean that the phenomenon does not exist.

To conclude, we should notice the presence of the following relationships:

$$\left| \delta_{4E} \right| \approx 0.5 \left| \delta_{1E} \right|$$

and:

$$\left| \delta_{5E} \right| \approx 2 \left| \delta_{2E} \right|$$

These relationships are, in some sense, evident, as the pairs of errors are caused by the same physical phenomena (i.e., the phase averaging along a measuring antenna and the amplitude averaging by it), and caused by these effects on identical antennas, but in different spatial configurations.

4.5 E-FIELD PROBE COMPARISON

A probe designated for wideband EMF measurements in the near field should meet the following requirements:

- The probe's antenna should be considerably smaller in size as compared to the minimal wavelength of the frequency range in which the probe is to work, and the size should not exceed (at first approximation) the minimal probe–source distance at which the measurement is to be performed.
- The transmittance of the probe should be as high and flat as possible within the measuring frequency band, while the magnitude of the transmittance outside of the band should not exceed that within the band.

and sometimes:

- The lower corner frequency should be as low as possible.

The two first requirements contradict one another, which creates the necessity of finding a compromise solution because, for instance, an increase in antenna sizes results in a sensitivity increase; simultaneously, however, the increase of the sizes results in an increase in errors caused by the mutual interaction of the antenna and the field source. As a result, there is a decrease in the permissible upper corner frequency of the probe equipped with the antenna, and the probe–source distance must be enlarged where the measurement is to be performed if we want the measurement accuracy to remain unchanged.

Both necessities of comparing probes of different parameters and then optimizing them require the introduction of a common measure that would take into account the three above-mentioned requirements. We will call this measure the *quality factor* of the probe (q), and we will define it in the form:

$$q = \frac{T(jf)}{f_1}$$

(4.49)

As an example of the proposed approach to the field probe comparison, we discuss the possible application of the measure to the evaluation of three different types of antennas, namely a thin, symmetrical dipole antenna, the same with a capacitive loaded top (double-T antenna), and a biconical antenna. We will assume the antenna length, h, and the detector input impedance (R_d, C_d) are the same for each antenna. This is because the magnitudes in Eq. (4.49) are functions of the antenna length h requirements and the detector input impedance. We will use the probe with the thin dipole antenna, whose quality factor is q_d as a reference. We will define the gain g_c of the capacitive loaded antenna resulting from the extension of the dipole with its capacitive loading, in relation to the probe with the dipole, in the following form:

$$g_c = \frac{q_c}{q_d}$$

(4.50)

where q_c is the quality factor of the probe with a capacitive loaded antenna.

Analogously, we will define the gain g_b of the probe with a biconical antenna:

$$g_b = \frac{q_b}{q_d}$$

(4.50a)

where q_b is the quality factor of the probe with the biconical antenna. Substituting Eqs. (4.14) and (4.15) for (4.49), we have:

$$q = 2\pi h_{eff} C_A R_d \tag{4.51}$$

The quality factor of the probe with a dipole antenna is:

$$q_d = 2\pi h^2 R_d \; \frac{10^{-12}}{3,6 \, (\ln h/a \, - \, 1,7)} \tag{4.52}$$

In the case of a double-T dipole antenna with a total length 2h, made of a conductor of diameter 2a and extended by a loading capacitance of length h_1 and conductor diameter $2a_1$, the antenna's electrical parameters are given by the following formulas:

Input capacitance C_i:

$$C_i = C_A + C_1 \tag{4.53}$$

where:

C_A = input capacitance of dipole without the capacitive extension

C_1 = equivalent capacitance of the extension

Effective length:

$$h_{eff} = h \; \frac{h + 2h_1}{h + h_1} \tag{4.54}$$

After substituting Eqs. (4.53) and (4.54) into Eqs. (4.49) and (4.50), we will have obtained the quality factor q_c and the gain g_c of the probe equipped with the capacitive extended symmetrical dipole:

$$q_c = q_d \; \frac{h + 2h_1}{h + h_1} \left(1 + \frac{h_1}{h} \; \frac{\ln h/a \, -1,7}{\ln h_1 / a_1 -1,7} \right) \tag{4.55}$$

and:

$$g_c = \frac{h+2h_1}{h+h_1}\left(1+\frac{h_1}{h}\,\frac{\ln h/a -1,7}{\ln h_1/a_1 -1,7}\right)$$

(4.56)

In the case of a biconical antenna of total length 2h and apex angle Θ_b, we can calculate its quality factor q_b and the gain g_b using the following formulas. This, respectively, gives the input capacitance C_i and the effective length h_{eff} of the short biconical antenna:

$$C_i = \frac{h\left(1+\dfrac{\ln 4}{2\ln \operatorname{ctg}\Theta_b/2}\right)}{3,6\,\ln\operatorname{ctg}\Theta_b/2}$$

(4.57)

$$h_{eff} \approx h\,(1+\sin\Theta_b/2)$$

(4.58)

The substitution of Eqs. (4.57) and (4.58) into Eqs. (4.49) and (4.50a) gives:

$$q_b = 2\pi h^2 R_d\left(1+\sin\Theta_b/2\right)\frac{1+\dfrac{\ln 4}{2\ln\operatorname{ctg}\Theta_b/2}}{3,6\,\ln\operatorname{ctg}\Theta_b/2}$$

(4.59)

and:

$$g_b = \frac{\left(1+\sin\Theta_b/2\right)\left(1-\dfrac{\ln 4}{2\ln\operatorname{ctg}\Theta_b/2}\right)\left(\ln\dfrac{h}{a}-1,7\right)}{\ln\operatorname{ctg}\Theta_b/2}$$

(4.60)

Table 4.1 shows calculated and measured magnitudes of the quality factor and gain of the probes with the mentioned antennas. Both the calculations and the measurements were performed for $C_d = C_m + C_f$ (where C_m = equivalent capacitance of the detector, including the probe's parasitic capacitances. For the set under investigation, it was estimated that $C_m \approx 1.8$ pF, while the capacitance of the detector itself was about 0.8 pF), and $C_f = 0$, 10, and 20 pF. The detector's equivalent resistance R_d was estimated as equal to about 2 MΩ. The results of measurements, as presented in the table, were obtained for the E-field

Table 4.1   Comparison of E-Field Probe Parameters with Different Antennas

Antenna	C_f [pF]	Theory T_0 mV/V/m	Theory f_l [kHz]	q	Experiment T_0 mV/V/m	Experiment f_l [kHz]	q	Gain g Theory	Gain g Experiment
Thin	0	12.0	34.6	3.45	11.6	21.2	5.47		
symmetrical	10	2.24	6.47		2.20	3.8	5.79		
dipole	20	1.23	3.57		1.21	2.11	5.73		
Dipole with	0	38.5	20.6	18.7	25.2	14.5	17.4	5.4	2.9
capacitive	10	10.7	5.74		5.51	3.52	15.7		
extension	20	6.20	3.33		3.13	1.18	16.6		
Biconical	0	37.3	20.2	18.5	26.5	15.1	17.5	5.4	3.2
	10	10.6	5.71		5.81	3.48	16.7		
	20	6.15	3.33		3.33	1.89	17.6		

probe in which different combinations of the same detection network were applied while filter capacitances C_f and antennas were replaced. For the calculations and the measurements, antennas of the following sizes were applied:

- A symmetrical dipole antenna, h = 5 cm, 2a = 1.0 mm
- A capacitive extended antenna, h = 5 cm, h_l = 3.5 cm printed on a substrate in the form of 5-mm wide strips
- A biconical antenna printed on a substrate in the form of two triangles where h = 5 cm and the triangle base = 3.5 cm

The comparison allows us to formulate several comments:

1. The measured magnitudes of transmittance are below those estimated, which is the result of using formulas that are valid for spatial structures for parameter estimation of flat structures.

2. For similar reasons, there is a discrepancy between the calculated and measured magnitudes of the quality factor and gain.

3. The measured lower corner frequencies are, in general, below those calculated; this allows us to suppose that the estimated equivalent resistance of the detector was undervalued.

4. The quality factor and gain of the probe with the capacitive extended dipole antenna and those of the biconical one are similar. However, because of the biconical antenna's lesser input impedance

variations versus frequency, the antenna may be accepted as more convenient for super-wideband measurements.

5. The presented method of E-field probe parameter comparison may also be applied as a comparison criterion for aspects other than the probe's antenna type, as, for instance, for comparison and optimization of other probe and meter designs.

4.6 COMMENTS AND CONCLUSIONS

The comparison of E-field probes has allowed us to formulate several conclusions regarding the optimization of antenna types used with them. The conclusions are, in some sense, evident, but it is important that the approach may be used by meter manufacturers in the planning and design stages as well as instrument users for purposes of evaluation and comparison of meters available on the market.

However, the most important thing here is the possibility of estimating measurement accuracy as a function of all of the accuracy-limiting factors discussed in the chapter, and by other factors that, for purposes of a particular required measurement, should be found, defined, estimated, and included so that the reader may independently evaluate measurement inaccuracy.

There is a possibility of almost arbitrarily shaping the frequency response of a wideband probe by the way of an appropriate choice of the corner frequencies of filters that are used to shape the response. For example, based on the grounds of permissible magnitudes of errors δ_{1E} and δ_{4E}, it allows remarkable limitation of the probe's sensitivity at frequencies outside the measurement band. On the other hand, the errors resulting from the accepted relations of R_0/h and h/λ determine both the properties of the probe with a selected type of measuring antenna and the applicability of the probe in near-field measurements. Consequently, we devoted most of our attention to their analysis.

The errors δ_{1E}, δ_{2E}, δ_{3E}, δ_{4E}, and δ_{5E} were estimated only for symmetrical dipole antennas. Asymmetrical antennas for near-field EMF measurements may be seen to be impractical based on the capacitive couplings of the antenna with surrounding objects and the role of the person taking the measurement as a "counterpoise" of the antenna. Consideration of other types of antennas (for example, capacitive extended dipoles, cylindrical, or biconical ones). This was omitted here as it would not reveal anything new or beyond the considerations presented.

The considerations included here allow precise estimation of the maximal magnitudes of measuring errors resulting from the size of the probe's antenna as well as from the distance between the antenna and the radiation source. As a result, we can estimate the maximal sizes of the measuring antenna or the inverse. This allows us to establish the minimal distance from a source at which the measurement will be loaded by an error of acceptable magnitude. It is especially important when studying commercially available instruments, because the information on how the measuring error increases when measurements are performed close to material bodies is not published by meter manufacturers. Their literature instead reveals much smaller errors, which are valid only in the far field.

The errors resulting from differences in the measured field phase along the antenna are identical for the longitudinal and the transversal field components. However, errors resulting from R_0/h ratio are maximal when the longitudinal component is of concern. Moreover, the transversal component of the electric field approaches zero near a perfectly conducting surface. Because of this, errors δ_{2E} and δ_{5E} were calculated for the longitudinal component but, during the component measurement, the antenna is placed perpendicularly to the surface of the source.

The influence of the conducting surface was analyzed when the antenna is located parallel to it (error δ_{3E}). Thus, the error is overestimated when the longitudinal component is measured and, as a result, it majorizes all possible errors resulting from the effect. This approach, however, was necessary, as measurements may be performed near complex radiating structures, containing a number of active and passive radiators, and it was used to show the domination of errors δ_{2E} and δ_{5E}. The role of these errors may be demonstrated in the calibration conditions in a TEM cell. When the probe is brought nearer to one of the cell walls, the meter's indications increase.

Errors δ_{1E} and δ_{4E} are systematic ones, and they may be precisely calculated for every measuring antenna if the frequency of the source is known. Errors δ_{2E}, δ_{3E}, and δ_{5E} are systematic as well, and they may be estimated when the conditions of the measurement are known (known parameters of the source, parameters of the measuring antenna, and the geometry of propagation). Usually, however, the parameters of the source, and as a result, the propagation geometry, may not be known *a priori*. Sometimes even the source itself may not be known before the measurements are made. As mentioned above, some parameters of the probes (meters) are not revealed by manufacturers. This leads to the necessity of assuming (especially when measurements of a surveying or monitoring character are performed) that these errors are accidental ones.

It may be assumed that the sizes of the measuring antenna are usually much smaller as compared to those of the radiation source. Thus, the source may be considered, in relation to the probe, as "an infinitely long line," generating in its vicinity the cylindrical wave which radial E-field component dominates in its proximity. In this case, the error should not exceed its magnitude estimated for $\alpha = 1$, and it may be recognized as the most probable error when field sources of practical importance are investigated.

We should note here that errors δ_{2E} and δ_{3E} are always positive, which causes measurements near material media to be overestimated in relation to standardized (meter calibration) conditions. It explains the increase in meter readouts when measurements are performed close to a primary or secondary source. From a purely metrological point of view, the incidence of an accidental error is of relatively low importance. However, in our field, where possible EM radiation hazards are of concern, the positive character of the error may be, in some sense, an advantage, as the real field strength should not exceed the measured one. That said, we should continue to make the effort to measure the E-field (and any other physical quantity) with the desired accuracy.

References

[1] T. Babij, H. Trzaska. Wideband properties of electric field probes. *Proc. 1975 IEEE Intl. EMC Symp.*, pp. 5BIa1–6, San Antonio, TX.

[2] R. W. P. King. *The theory of linear antennas.* Harvard Univ. Press, 1956.

[3] M. Kanda. Analytical and numerical techniques for analyzing an electrically short dipole with a nonlinear load. *IEEE Trans.* vol. AP-48, No. 1/1980, pp.437–442.

[4] G. Z. Ajzenberg. *Shortwave antennas* (in Russian). Sviazizdat, Moscow 1962.

[5] M. Kanda, F. X. Ries. A Broad-Band Isotropic Real-Time Electric-Field Sensor (BIRES) Using Resistively Loaded Dipoles, *IEEE Trans,* vol. EMC-23, No. 3/1981, pp. 122–132.

[6] M. Kanda, L. D. Driver. An Isotropic Electric-Field Probe with Tapered Resistive Dipoles for Broad-Band Use, 100 kHz to 18 GHz, *IEEE Trans.* Vol. MTT-35, No. 2/1987, pp. 124–130.

[7] P. Bienkowski. Electromagnetic fields measurements—methods and accuracy estimation, *Studies in Applied Electromagnetics and Mechanics.* Vol. 29, IOS Press. 2008, pp. 229–237.

[8] P. Bienkowski. Parameters of wideband electromagnetic field sensors and possibilities of their modification, *Applied Electronics 2005. Proc. International Conference,* Pilsen, 2005.s. 41–44.

Chapter 5

Magnetic Field Measurement

This chapter discusses properties of the probes for RF magnetic field measurements and in particular, the factors limiting the measurement accuracy. Our considerations are limited to a probe consisting of a circular loop antenna loaded with a detector of a shaped frequency response. However, the majority of results are fully applicable for Hall-cell probes, magneto-optic probes, those with a magneto-diode, and for other designs, especially when considering averaging of the measured field upon the surface of the measuring antenna (probe).

Many of the considerations are similar in character to those presented in Chapter 4, and some are concerned with the magnetic field measurement specificity.

5.1 THE SIZES OF THE MEASURING ANTENNA

Via the analogy used in Section 4.1, it is possible here to relate to the measuring antenna size limitation to the error of a quasi-point value of the magnetic field measurement. It concerns the electrical sizes of the antenna in particular. The discussion leads to conclusions similar to those presented in Section 4.1 by substituting πr (where r is radius of a circular loop antenna) for h. However, a more rigorous limitation of the loop antenna sizes results from the so-called "antenna effect"—in other words, from simultaneous sensitivity of the loop antenna (especially a non-screened one) to the electric field as well.

The current in a load of the loop antenna I_l may be expressed as:

$$I_l = \lambda K_H cB \pm \lambda K_E E = I_H \pm I_E \tag{5.1}$$

where:

K_H, K_E = sensitivity of an unloaded loop to the magnetic and electric field respectively,

    c = in general, the velocity of propagation, and here the velocity of propagation in the vacuum:

$$c = \frac{1}{\sqrt{\varepsilon_0 \mu_0}}$$

I_H, I_E = components of the current in the antenna load proportional to the magnetic and the electric field.

Equation (5.1) was introduced for the plane wave and positioning of the loop in relation to the field components such that I_H is of maximal value while I_E is of maximal or minimal value. The resulting measurement error due to the presence of the antenna effect (δ_{1H}) we will define as:

$$\delta_{1H} = \frac{\left| I_H - I_E \right|}{\left| I_H \right|} \approx \frac{\left| K_E \right|}{\left| K_H \right|} \frac{\left| Z_a \right|}{Z_0} \tag{5.2}$$

where:

    Z_a = wave impedance:

$$Z_a = \frac{E}{H} \tag{5.3}$$

independent of the source type $\left| Z_a \right| \in \, <0, \infty>$

Z_0 = intrinsic impedance of free space:

$$Z_0 = \sqrt{\frac{\varepsilon_0}{\mu_0}} \approx 120\pi \ [\Omega]$$

E, H = complex amplitudes of E and H vector.

In accordance with King [1]:

$$\frac{K_E}{K_H} = -j2\pi \, \frac{2r_0}{\lambda} \tag{5.4}$$

To reduce the significance of the antenna effect, a symmetry is exploited and/or screening of the loop is applied. For the same purpose, double- or, in general, multiple-loaded loop antennas were proposed. An example of a multiple-loaded loop application will be presented in Chapter 6. The error δ_{1H} response as a function of kr_0 for several values of Z_a is plotted in Fig. 5.1.

Equations (2.21–2.23) as well as (5.2) allow us to write, for the elemental magnetic dipole:

$$\lim_{R_0 \to 0} \delta_{1H} \to \infty$$

$$(5.5)$$

In light of this, the magnetic field measurement close to a short electric antenna (where the electric field is the "dominant" one) may lose its meaning. A radical way of reducing the antenna effect is to use an electrostatic screen. However, this is not an ideal solution because a simultaneous increase in the antenna capacitance results in a decrease of both the antenna's resonant frequency and the sensitivity of the probe with such an antenna. Analogous to the considerations in Section 4.3, and based on Eqs. (2.28–2.30), it is possible to say, for an infinitely long line source, if $z \to 0$ and $\rho \to 0$, then $|Z_a| \le Z_0$. However, for $z \to h$, the inequality should be reversed. As a test of the magnitude of the measurement error resulting from the antenna effect, it is advisable to use a criterion based upon an investigation of the asymmetry (σ) of the directional pattern of the measuring antenna (probe). The criterion is simple, and it may be used for both for single-loop probes and omnidirectional ones:

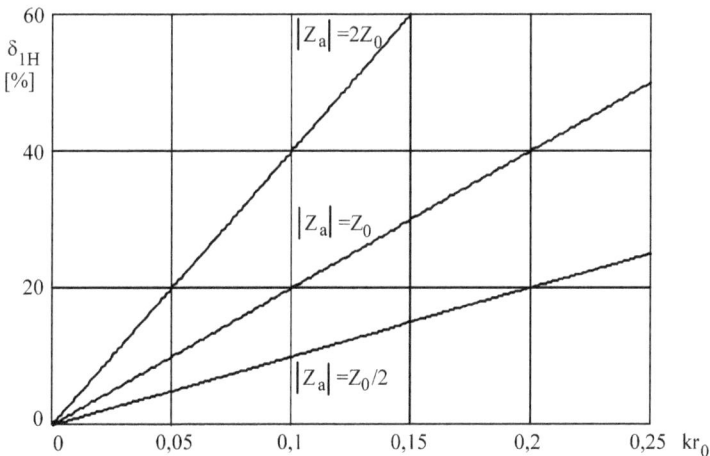

Figure 5.1   The error δ_{1H} response as a function of kr_0.

$$\sigma = \frac{\alpha_1 - \alpha_2}{\alpha_1 + \alpha_2} = \left| \delta_{1H} \right|_{max}$$

(5.6)

where:

α_1, α_2 = indications of the magnetic field meter responding to both maxima of its directional pattern.

5.2 FREQUENCY RESPONSE OF THE MAGNETIC FIELD PROBE

The effective length h_{eff} of a loop antenna fulfilling the condition $kr_0 \ll 1$ is directly proportional to the frequency. In order to obtain a frequency-independent response of the magnetic field probe below the resonant frequency of the antenna, the antenna is loaded with a four-terminal network whose frequency response is proportional to λ.

An equivalent network of the magnetic field probe at low frequencies, composed of an antenna and a response-correcting network, is shown in Fig. 5.2.

The transmittance of the probe is given by:

$$T(jf) = \frac{-2\partial nSi_0 f}{\left(1 - \frac{f^2}{f_1^2}\right)\left(1 + \frac{f_3}{f_3}\right) - \frac{f^2 f_4}{f_1^2 f_3} + j\left[\left(1 - \frac{f^2}{f_1^2}\right)\frac{f}{f_3} + \frac{ff_4}{f_1^2}\left(1 + \frac{f_3}{f_3}\right)\right]}$$

(5.7)

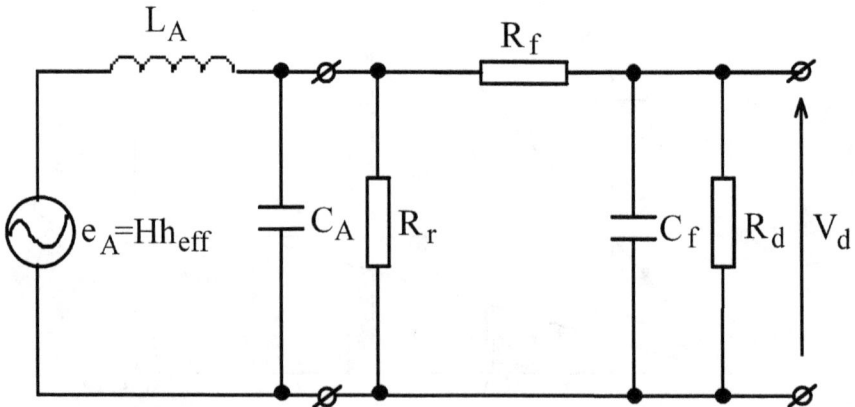

Figure 5.2   Equivalent network of the magnetic field probe.

where:

$$f_1 = \frac{1}{2\pi\sqrt{L_A C_A}}$$

$$f_2 = \frac{1}{2\pi\, R_d C_f}$$

$$f_2 = \frac{1}{2\pi\, R_d C_f}$$

$$f_3 = \frac{1}{2\pi\, R_f C_f}$$

$$f_4 = \frac{1}{2\pi\, R_r C_A}$$

$$h_{eff} = 2\pi\, nS\mu_0 f \tag{5.8}$$

n = number of turns,

S = the surface area of the loop antenna.

To achieve an optimal shape of the probe's frequency response, it is convenient to assume:

$$R_r = 2\pi f_1 L_A$$

In other words, $f_1 = f_4$.

On the grounds of Eq. (5.7), it is possible to determine the following frequency segments of the magnetic field probe:

1. Low frequencies where:

$$\left| T(jf) \right| = \frac{2\pi nS\mu_0 R_f f}{R_f + R_d} \tag{5.9}$$

2. Medium frequencies for which $f_2 \ll f_3$, which is always fulfilled, $T_0(jf) \neq f(f)$, and the transmittance is given by:

$$T_0(jf) = \frac{2\pi nS\mu_0}{R_f C_f} \tag{5.10}$$

We may take note of the lower corner frequency (f_l) of the frequency band, in which the transmittance decreases by 3 dB, $f_l = f_3$, whereas the upper corner frequency of the probe $f_u \approx 1.5\ f_1$. However, the relation $f_4 \approx 2f_1$ is usually applied, which means that the transmittance has a certain maximum for $f = f_1$. Sometimes, in order to eliminate the maximum, another RC low-pass filter (or two) of $f_g \approx 0.5\ f_1$ is applied, then $f_u = f_g$.

3. High frequencies at which:

$$\left| T(jf) \right| = \frac{2\pi nS\mu_0 f_1^2 f_3}{f^2}$$

(5.11)

The calculated magnetic field probe's frequency response, whose equivalent diagram is shown in Fig. 5.2, for $f_4 = 2f_1$, is shown in Fig. 5.3 (continuous line). The dashed line generalizes the measured frequency response of the probe equipped with an additional double RC low-pass filter of $f_g = 0.5\ f_1$. Both curves were normalized in relation to T_0.

An accurate analysis of the magnetic field probe equipped with a triple RC low-pass filter has been performed. The first segment of the filter allows us to obtain the required lower corner frequency. The two-segment filter lets us attenuate signals at the antenna's self-resonant frequency as well as eliminate any other maxima and reduce the probe's sensitivity above its upper corner frequency. Especially important here is the maximal attenuation at frequency at which the antenna works as a "folded dipole" and its input impedance is relatively low. The analysis takes into account the HF equivalent network. It was

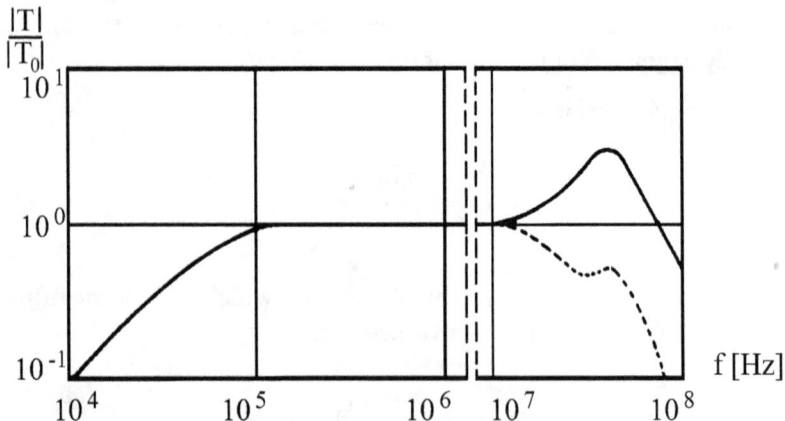

Figure 5.3 Transmittance of the wideband magnetic field probe.

found that the local maxima of transmittance may appear, especially at frequencies corresponding to the resonance of the antenna's self-inductance and the parasitic reactances of the probe. However, the magnitude of the transmittance at these maxima does not exceed −20 dB in relation to the transmittance within the measuring band. The estimations and experiments have fully confirmed that the use of additional RC filters is indispensable here. Without them, the transmittance of a probe designed to work within a frequency range up to 20 (30) MHz, at frequencies above about 100 MHz, may be beyond any control; while fringes at these frequencies appear during measurements, they may lead to errors that totally invalidate the results. The comment regarding the necessity of assuring a controllable run of the frequency response of any probe outside the measuring band, and especially at the highest frequencies, is of primary importance. This response should be known to a person performing measurements.

5.3 DIRECTIONAL PATTERN ALTERNATIONS

The solution of the integral equation, which describes the current intensity along the circular loop antenna shown in Fig. 5.4, may be presented in the form [2]:

$$I(\varphi) = \frac{-jV_0}{Z_0 \pi} \left(\frac{1}{a_0} + 2 \sum_{n=1}^{\infty} \frac{\cos n\varphi}{a_n} \right)$$

$$(5.12)$$

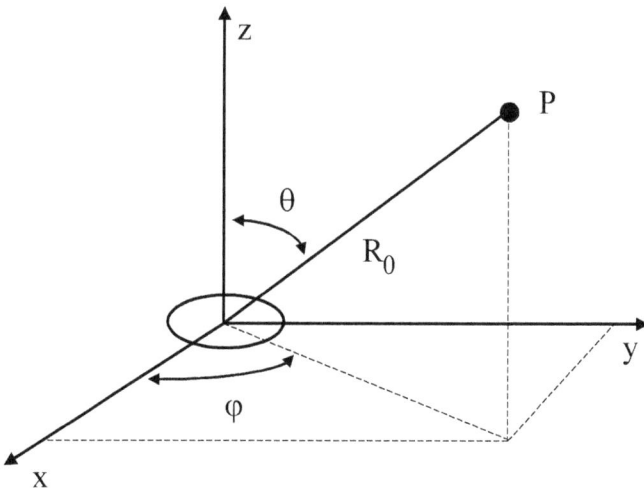

Figure 5.4 A circular loop in a coordinate system.

where:

φ = the central angle of the antenna,

V_0 = the voltage of the antenna exciting source,

a_n = a coefficient of the Fourier series expansion.

After substituting Eq. (5.12) into Eq. (2.8) and then into Eqs. (2.13) and (2.14), we will have spatial components of the magnetic field in the far field. For a loop antenna, located in the coordinate system as shown in Fig. 5.4, which fulfills the condition $kr_0 \leq 0.2$, the dominant component is $H_{\theta 0}$ (where index zero indicates zeroth order of approximation). This component corresponds to taking into account only term a_0 in the series (5.12), which indicates a uniform current distribution along the antenna (characteristic to the elemental dipole), then:

$$\left[H_{\theta 0} \right]_{xy} = \frac{j\omega\varepsilon_0 V_0 r_0}{2R_0} J_1(kr_0 \sin\theta) = AJ_1(kr_0 \sin\theta)$$

(5.13)

If two terms are taken into consideration in the series (5.12), i.e., a_0 and a_1, then the approach is equivalent to the first order of approximation that is equivalent to the cosine current distribution. The approximation is true for $kr_0 \leq 1$, then:

$$\left[H_{\theta 1} \right]_{xy} = A_1 J_1(kr_0 \sin\theta) \cos\phi$$

(5.14)

$$\left[H_{\phi 1} \right]_{xy} = A_1 \frac{J_1(kr_0 \sin\theta)}{kr_0 \sin\theta} \cos\theta \sin\phi$$

(5.15)

where:

A, A_1 = constants dependent on R_0

J_1 = Bessel function of the first order

If three identical loop antennas are placed in three mutually perpendicular planes and their directional patterns are summed by squaring, which is equivalent to the load of each antenna with a square-law detector, then we will obtain a magnetic field probe with a spherical radiation pattern (omnidirectional probe). This is shown below. It is necessary to remember here, however, that although the radiation pattern is calculated in specifically for a transmitting antenna, by making the use of the reciprocity theorem, the results can be applied for a receiving antenna. However, such an antenna may function only as the receiving one.

Let's consider the radiation pattern of a system consisting of three mutually perpendicular loop antennas for the zeroth-order approximation of the current distribution in them. We will define the radiation pattern f_{xy} of a loop antenna placed upon the xy plane of the Cartesian coordinate system in the form:

$$f_{xy} = \frac{\left[H_{\theta 0} \right]_{xy}}{\left[H_{\theta 0} \right]_{xy\ max}} = \frac{J_1(kr_0 \sin \theta)}{kr_0} \tag{5.16}$$

and those of the antennas on the planes xz and yz in the forms:

$$f_{xz} = \frac{J_1(kr_0 \sqrt{1 - \sin^2\theta \sin^2 \varphi}}{kr_0} \tag{5.17}$$

and:

$$f_{yz} = \frac{J_1(kr_0 \sqrt{1 + \sin^2\theta \cos^2 \varphi}}{kr_0} \tag{5.18}$$

Thus, for $x \ll 1$ and $J_1(x) \approx x$, and by analogy with Eq. (4.42), for antennas loaded with the square-law detectors, we will have:

$$F(\theta,\varphi) = f_{xy}^2 + f_{xz}^2 + f_{yz}^2 = 2 \tag{5.19}$$

The directional pattern given by Eq. (5.19) is direction independent, with no regard to the coordinates applied. It is true for $kr_0 \ll 1$. Such a case, however, is extremely interesting to us, as it results from considerations presented in Section 5.1.

For antennas that do not fulfill the condition $kr_0 \ll 1$, it would be necessary to repeat the estimations. For instance, consider the first approximation, for which the radiated field is given by Eqs. (5.14) and (5.15), and to accept a definition of the pattern irregularities identical to that given by Eq. (4.43) is desired. If an antenna fulfills the condition $2\pi r_0 \leq \lambda$, the results of the pattern's nonuniformity estimations are identical to those presented in Fig. 4.4, when the substitution $h = \pi r_0$ is used.

Similarly, as in Section 4.4, during estimations of the spherical directional pattern with dipole antennas, when the spherical pattern of three mutually perpendicular loop antennas is synthesized, the amplitude variations on the surface of the antennas were found to be the most influential. It was shown that if $kr_0 \leq 0.2$, the role of the phase variation upon the shape of the pattern is negligible. Thus, the sizes of the loop antenna in relation to the wavelength are limited, first of all, by the permissible magnitude of the error δ_{1H}. Again, per analogy with the pattern deformation δ_{5E} [Eq. (4.48)], it is possible to estimate the irregularities of the spherical pattern of a system of three loop antennas as a function of r_0/R_0. The estimation may conclude that the nonuniformity is approximately twice as much as the error δ_{2H}. A similar conclusion has already been formulated for the electric field probe.

As far as the shape of the spherical directional pattern is considered and the above errors are discussed, it is worth recalling that all of this discussion, as in the case of E-field and H-field probes, was performed with one assumption: all the omnidirectional probes were placed in the center of the coordinate system. The assumption was correct, and results of the estimations and final comments are correct as well. However, in practice, neither E-field nor H-field probes are always placed symmetrically in the relation to the coordinate system [3]. An example illustrating the above is shown in Fig. 5.5. A design of the "symmetrical" probe 3AH of the MEH type meter is shown in Fig. 5.5a, while the loops' location in an "unsymmetrical" EHP-50 type meter is shown in Fig. 5.5b, with the case removed.

As an example, here we will analyze an error due to asymmetry in a probe's construction as shown in Fig. 5.5b. The probe consists of three

(a)   (b)

Figure 5.5   (a) Symmetrical and (b) unsymmetrical H-field probe designs.

loops placed at the side walls of a cube of external dimensions $d \approx 8$ cm; thus, the diameter of any loop is slightly less than d. Figure 5.6 presents the results of estimations of the additional H-field measurement error when the probe is placed on the z axis of the Cartesian coordinate system (Fig. 5.4) at a distance R_0 between the source, in the form of a loop antenna of radius $r_0 \approx d/2$ placed in the xy plane of the coordinate system, and the nearest wall of the cube. A probe's loop placed at a plane parallel to plane xy is located at opposite side of the cube in relation to the source, i.e., at a distance $D \approx R_0 + d$.

These estimations were performed using Eq. (5.6). The results allow several comments:

- The estimations presented were performed for the selected type of H-field probe, and the results are valid only for this type of probe and similar exposure conditions.

- The assumed configuration reflects standard measurements conditions while the probe–source distance is defined as above. It may be observed that, under the analyzed conditions ($D \approx R_0 + d$), the error is of a negative character. In the opposite case ($D \approx R_0$), it would be of a positive character, as a result of calibration conditions, which requires a relation to the geometrical center of the probe.

- The error is specific to any type of probe of unsymmetrical construction, and it should be estimated individually for any probe and measured field situation.

- A similar concept of unsymmetrical omnidirectional probe construction is applied in E-field probes, in H-field probes with Hall cells,

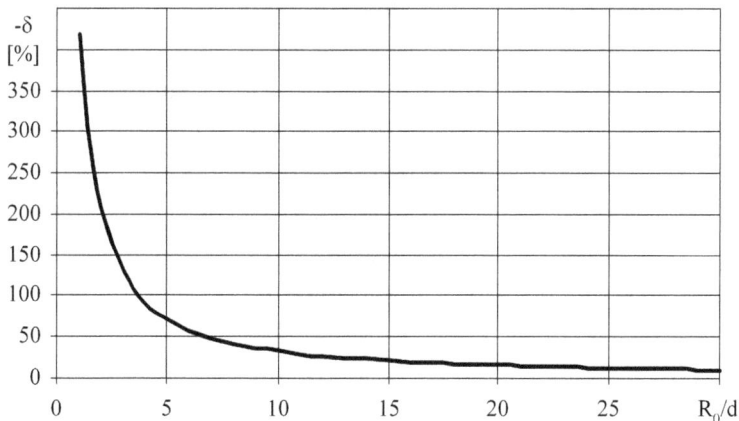

Figure 5.6 Estimated run of error versus R_0/d.

and such. However, their sizes are relatively small and, as a result, the error level is comparatively smaller.

- If there are any doubts about the presence (and role) of the discussed error, the procedure presented in Section 5.1, based on Eq. (5.6), may be helpful.

5.4 MEASUREMENT ACCURACY VS. ANTENNA DISTANCE FROM RADIATION SOURCE

5.4.1 Influence of the Source Structure on Accuracy

The electromotive force e_H induced by the field in a loop antenna of surface area S is given by Faraday's law:

$$e_H = -j2\pi n\mu_0 f \int_S \mathbf{H} \cdot \mathbf{dS}a$$

$$(5.20)$$

If the radiation source is the magnetic elemental dipole located at point P (R_0, θ, φ) at distance R_0 from the center of the circular loop antenna, as shown in Fig. 5.4, and the dipole sizes are much less than R_0, then, making the use of Eq. (2.27), we may write:

$$e_H = -j2\pi n\mu_0 f C_2(\alpha) \int_S \frac{\exp(-jkR)}{R^\alpha} \cos(\mathbf{H}, \mathbf{dS}) \, dS$$

$$(5.21)$$

where:

R = distance between the integration point and the field source:

$$R = \sqrt{R_0^2 + x^2 + y^2 - R_0 \sin\theta \, (x\cos\varphi + y\sin\varphi)}$$

$$(5.22)$$

and x, y = coordinates of the integration point: x, y ∈ S.

The maximal phase and amplitude shifts of the field on the surface of an antenna will appear for the transversal field component when **H** is parallel to **S** and $\theta = \pi/2$. If we assume, for instance, $\varphi = 0$, we will have:

$$e_H = -j \frac{h_{eff}}{S} C_2(\alpha) \int_S \frac{\exp(-jkR)}{R^\alpha} \, dxdy$$

$$(5.23)$$

and:

$$R = \sqrt{\left(R_0 - x\right)^2 + y^2}$$

(5.24)

where: h_{eff} = effective length given by Eq. (5.8).

The presence of the antenna effect limits loop antenna sizes much more rigorously, as compared to the field averaging over the surface of the antenna and caused by the phase differences on the surface. It allows, similarly as in Section 4.2, the assumption in Eq. (5.23) of:

$$\exp\left(-jkR\right) \approx \exp\left(-jkR_0\right)$$

Then:

$$e_H = -j\, \frac{h_{eff}}{S}\, C_2(\alpha)\, \exp\left(-jkR_0\right) \int_S \frac{1}{R^\alpha}\, dx\,dy$$

(5.25)

In the case of the homogeneous field, the emf e'_H is:

$$e'_H = -jh_{eff} C_2(\alpha)\, \exp\left(-jkR_0\right) R_0^{-\alpha}$$

(5.26)

In order to introduce a uniform reference system (standardization), it is indispensable to assume, as in other presented examples, that the probe's calibration is performed in the homogeneous field whose radius of curvature is equal to infinity, i.e., a plane wave. As a result, when the magnetic field measurements are performed in close proximity to a source, a δ_{2H} error will appear, which results from averaging the field of finite curvature on the surface of the probe. The error can be defined in the form:

$$\delta_{2H} = \frac{e_H - e'_H}{e_H}$$

(5.27)

After substituting Eqs. (5.25) and (5.26) into (5.27) we will have:

For $\alpha = 3$,

$$\delta_{2H} = 1 - \left\{ \frac{4}{\pi k^2} \left[\frac{1}{1 - k^2}\, E(k) - K(k) \right] \right\}^{-1}$$

(5.28)

where:

E(k), K(k) = elliptic integrals of the first- and second-order and argument k:

$$k = r_0/R_0$$

For α = 2

$$\delta_{2H} = 1 - k^2 \left[\ln \frac{1}{1 - k^2} \right]^{-1}$$

(5.29)

where:

$k = r_0/R_0$, and

For α = 1

$$\delta_{2H} = 1 - \frac{\pi r_0}{2R_0} \left[\left(\frac{R_0}{r_0} + 1 \right) E(k) - \left(\frac{R_0}{r_0} - 1 \right) K(k) \right]^{-1}$$

(5.30)

where:

E(k), K(k) = elliptic integrals of the first and second order and argument k:

$$k = \frac{2 \sqrt{R_0 r_0}}{R_0 + r_0}$$

of course, for α = 0 $\delta_{2H} \equiv 0$.

Calculated values of the error δ_{2H} are plotted in Fig. 5.7.

5.4.2   The Role of Antenna Input Impedance Variations

The transmittance of the magnetic field probe, described in Section 5.2, is independent of the input impedance of the probe's antenna, which is derived from Eq. (5.10). Because of this, consideration of the antenna's input impedance changes, in proximity to a conducting medium, could be disregarded. However, attention should be focused on the fact that the overall measurement resulting from the measured field averaging

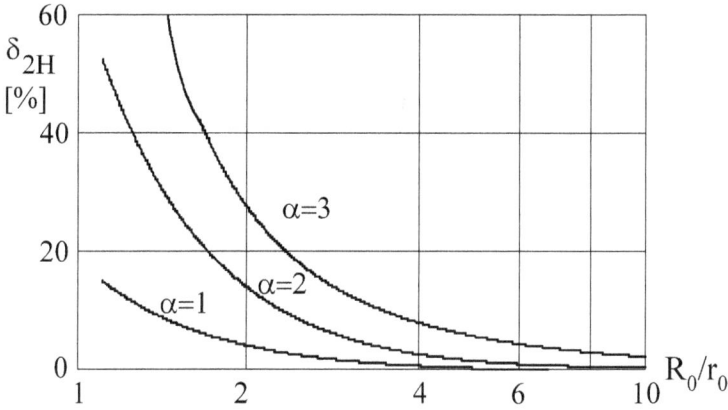

Figure 5.7   Error δ_{2H} as a function of R_0/r_0.

(especially in the near field), which is a function of r_0/R_0, for the considered probe design, is only a function of the structure of the radiation source (Fig. 5.7) and that the character of the error is always positive.

Keeping in mind that the considerations of Section 5.5 related to a somewhat different magnetic field probe design, as compared to that presented in Fig. 5.2, we will briefly review estimation of the loop antenna input impedance variations as a function of distance to a conducting medium. The medium, much as before, is infinitely large and perfectly conducting. The presence of such a medium may be replaced by a mirror reflection of the antenna under consideration.

The mutual inductance, M, of two identical loop antennas, collinearly placed at distance $2R_0$ is given by the ratio of the magnetic flux of the first antenna flowing thorough the surface area of the other S_2 to the current I_1, which the flux generates [4]:

$$M = \frac{\int\limits_{S_2} B_1 \, dS_2}{I_1}$$

(5.31)

After integration it becomes:

$$M = \mu_0 r_0 k \left[\left(\frac{2}{k} - k \right) K(k) - \frac{2}{k} E(k) \right]$$

(5.32)

where $E(k)$ and $K(k)$ = elliptic integrals of the first and second order and argument k:

$$k = \left[\left(\frac{R_0}{r_0} \right)^2 + 1 \right]^{-2}$$

For $R_0 \to 0$ $M \to L_A$, then:

$$L_A = \mu_0 \ (2r_0 - a) \left[\left(1 - \frac{k^2}{2} \right) K(k) - E(k) \right]$$

(5.33)

where $E(k)$ and $K(k)$ = elliptic integrals of the first and second order and argument k:

$$k = \frac{2\sqrt{r_0(r_0 - a)}}{2r_0 - a}$$

$2a$ = diameter of the conductor applied for the antenna windings; if $a/r_0 \to 0$, then:

$$L_A = \mu_0 r_0 \left[\ln \left(\frac{8r_0}{a} \right) - 2 \right]$$

(5.34)

Other approaches may yield similar results [5]. For the first approximation of the current distribution along the antenna, using Eq. (5.12), the input impedance of the antenna Z_A may be given in the form:

$$Z_A = \frac{j\pi Z_0 a_0 a_1}{2a_0 + a_1}$$

(5.35)

If we assume *a priori* that it is possible to construct a magnetic field probe whose transmittance within the measuring band is given by:

$$T(jf) = \frac{A}{L_A}$$

(5.36)

where A = a constant, then the measuring error, caused by the variations of the probe's (antenna) input impedance changes due to its parallel location at distance R_0 from a conducting plate as defined above, will be:

$$\delta_{3H} = \frac{M}{L_A}$$

(5.37)

The calculated values of error δ_{3H}, in the function of R_0/r_0, for several r_0/a ratios are shown in Fig. 5.8. It may be seen from Figs. 5.7 and 5.8 that errors δ_{2H} and δ_{3H} are of a positive sign. Thus, the resultant error is also positive one. It gives a non-underrated value of measured field strength and may be, in some sense, advantageous when human safety or environmental protection is of concern.

5.5 MAGNETIC FIELD PROBE WITH A LOOP WORKING ABOVE ITS SELF-RESONANT FREQUENCY

Figure 5.1 showed an equivalent circuit of the magnetic field probe in which the loop antenna was followed by an RC low-pass filter in order, to achieve the frequency independent transmittance within a certain frequency band. This solution is inconvenient, especially when detectors of low input impedance (e.g., thermistor or thermocouple ones) are to be used. The latter have several advantages (more univocal square-law dynamic characteristics, less susceptibility to interference, non-immune to static electricity).

As an equivalent alternative to the magnetic field probe, discussed in Section 5.2, we proposed the use of the properties of an electrically small loop antenna working at frequencies above its self-resonance (or

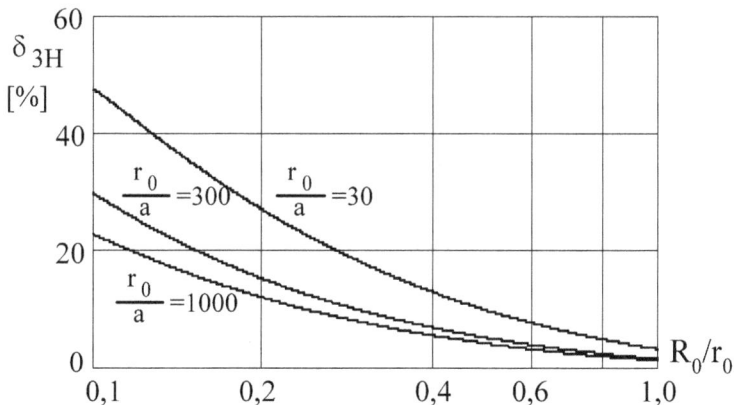

Figure 5.8 Error δ_{3H} as a function of R_0/r_0.

more precisely stated, the antenna has many resonant frequencies; here we consider the resonance of the small antenna inductance and parasitic capacitances of the probe; i.e., self-capacitance of the antenna eventually increased by an electrostatic screen, input capacitance of the antenna load, and the capacitance of the probe assembly). While considering the properties of such a probe, we will make use of the formula defining the effective length of the loop antenna with an approximation better than that of the zeroth order [Eq. (5.8)]. Based on Eq. (5.12), and using the reciprocity theorem, it is possible to estimate the effective length. Taking into account the phase distribution on the surface S of the antenna, its effective length may be expressed in the form:

$$h_{eff} = 2\pi n \mu_0 fS \left[1 - 2^{-3}(kr_0)^2 + \frac{2^{-6}}{3}(kr_0)^4 + \cdots \right]$$

$$(5.38)$$

It may be noted that the zeroth-order approximation of the loop antenna's effective length given by Eq. (5.8) corresponds to the first term of the series in Eq. (5.38). The transmittance of the magnetic field probe, composed of a loop antenna loaded with a serial connected RC filter (Fig. 5.9), will be defined as the ratio of the current I flowing in the loading resistance (and in the antenna) R_S to the intensity of the measured magnetic field H:

$$T(jf) = \frac{I}{H} = \frac{h_{eff}}{Z_s}$$

$$(5.39)$$

where Z_s = the equivalent impedance of the serial connection of the antenna input impedance Z_A and R_S and C_S.

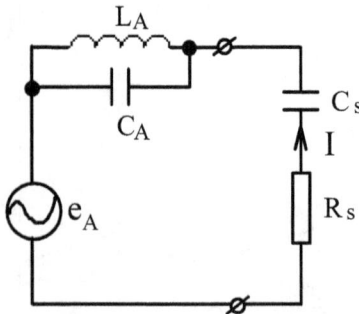

Figure 5.9 Equivalent network for the medium frequency range of an H-field probe.

Equation (5.39) is convenient when the properties of the probe are being considered. However, to obtain results that would be comparable to those obtained on the grounds of Eq. (5.7), it is necessary to multiply Eq. (5.36) by R_s. Then, substituting into Eq. (5.39) h_{eff}, given by Eq. (5.38), and Z_A, given by Eq. (5.35), we obtain the transmittance for the medium frequency band:

$$T_0(jf) = \frac{jnS\mu_0R_s}{L_A}$$

(5.40)

We may notice that Eq. (5.40) is analogous to Eq. (5.36).

For a maximally flat frequency response, i.e., for:

$$R_s = \sqrt{L_A / C_s}$$

the corner frequencies (−3 dB) of the medium frequency band are:

1. The lower corner frequency:

$$f_l = 0.62\, f_2$$

(5.41)

2. The upper corner frequency:

$$f_u = 0.5\, f_l$$

(5.42)

where:

$$f_l = \frac{1}{2\pi\sqrt{L_A C_A}}$$

(5.43)

$$f_2 = \frac{1}{2\pi\sqrt{L_A(C_A + C_s)}} \approx \frac{1}{2\pi\sqrt{L_A C_s}}$$

(5.44)

$$f_3 = \frac{1}{2\pi R_s C_s}$$

(5.45)

In this estimation, the above-mentioned effects, resulting from the resonance of the antenna as a folded dipole, have not been taken into account. Similar to the case of the antenna working below its self-resonant frequency, it requires the use of additional protection against unwanted increases in the probe's sensitivity at frequencies above the measuring band. Figure 5.10 shows plotted results of simplified calculations of the transmittance of a magnetic field probe while $f_3/f_1 = 1$, 3, and 10. The calculations made it possible to optimize the run of the transmittance in the low-frequency range. On the other hand, Fig. 5.11 shows the frequency response of the probe measured for $f_3 = f_1$. In the frequency response of the probe, without an electrostatic screen, an uncontrolled increase in the probe's sensitivity should be noticed. This can be explained as the approach to the frequency range in which the mentioned dipole resonances should be taken into consideration. The curves presented in Figs. 5.10 and 5.11 were normalized in relation to T_0.

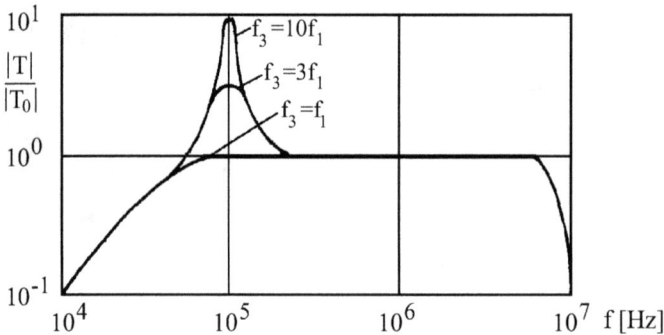

Figure 5.10   Estimated frequency response of the magnetic field probe.

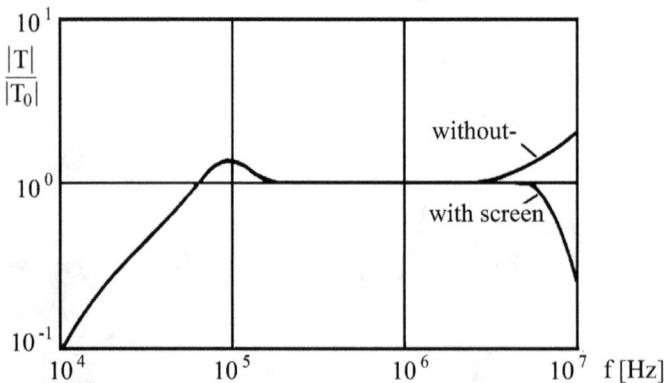

Figure 5.11   Measured frequency responses of the H-field probe.

In order to evaluate the applicability of a selected type of magnetic field probe, we defined, per analogy with considerations related to the electric field probes and presented in Chapter 4, the measure of the probe's quality in the form:

$$q = T_0(jf) \, \frac{f_u}{f_1}$$

(5.46)

By calculating the quality factor q and, similar to the previously defined gain factor of the magnetic field probes as presented in Sections 5.2 and 5.5, it is possible to show that, when loop antennas of similar parameters are used in the probe, working at frequencies both below and above the antenna self-resonance, the parameters of both probes are equivalent.

5.6 COMMENTS AND CONCLUSIONS

The maximal sizes of the magnetic field probe in which the loop antenna serves as the field sensor, for which the magnitude of error δ_{1H} is acceptable, may be estimated using the curves shown in Fig. 5.1. However, it may be necessary to use probes with different diameter antennas for measurements performed near different radiation sources. This is because the error is a function of E/H, which is a characteristic of the sources. This is an important issue, especially when known field sources are being investigated. Unfortunately, in the majority of cases, the nature of the source may be unknown. This would suggest the care and an assumption *a priori* that maximal errors may appear. For the purpose of determining the magnitude of error δ_{1H} and its importance evaluation, the criterion given by Eq. (5.6) may be useful. As mentioned above, the loop antenna size limitation resulting from the presence of the antenna effect is much more rigorous as compared to that resulting from the phase averaging on the antenna surface.

The accepted permissible magnitude of the error δ_{1H} is a starting point for evaluating the upper corner frequency and then the design of the filters shaping the detector's frequency response. Without regard to the assumed concept of the probe's operation (below or above self-resonance), the upper corner frequencies for identical antennas are similar. When the sizes of an antenna are known, it is possible to estimate the errors δ_{2H} and δ_{3H} as functions of R_0/r_0. In the case of the above-resonance probe, it is necessary to take into account both errors whereas, in the case of below-resonance probe, the variations of the antenna's input impedance do not have a remarkable influence on measurement accuracy.

Similar to the considerations presented in Chapter 4, for errors estimations in Chapter 5, the most disadvantageous measurement conditions have been assumed, which has led to the maximal error calculations. In order to find the maximal value of the error δ_{2H}, the error was calculated for a situation equivalent to the longitudinal field component. The situation exists, for instance, near a heating coil of an induction furnace where the spatial field variation allows the assumption $\alpha = 3$. However, if the magnetic fields associated with HF currents in an "infinitely long cable" are measured, the calculation gives much more realistic results, even for $\alpha = 1$. It results from the occurrence and domination of the tangential field component near the surface of the conductor; this is an important difference between the electric and magnetic fields, and it results in a difference in the accuracy of both components' measurements.

The error δ_{3H} was calculated for parallel positioning of the antenna in relation to a conducting surface. As a result, during the transverse component of the field measurement, the value of the error is overestimated, and it majorizes all possible values of the error. Such an approach is consequently applied in our elaboration, and it is supported by the possibility of performing measurements near primary and/or secondary sources of a complex and unknown *a priori* structure.

In the case of an indeterminate field source, which occurs in applications specific to our considerations, we will assume the errors δ_{2H} and δ_{3H} to be accidental ones. Simultaneously, the use of small antennas (in the electric and geometric sense) allows the supposition that the most probable error δ_{2H} is that estimated for $\alpha = 1$.

Based on the presented analyses, it is possible to attempt an estimation of the resultant error of the magnetic field measurement using one of the probe's described versions or another design (e.g., a probe with a Hall cell). The situation may be seen as simple enough that precise determination of any discussed partial errors is analytically possible. Moreover, this creates a temptation to increase the accuracy of the field measurement via the individual measurement accuracy estimation, or at least making use of analytically determined correction factors while the final measurement results are completed. To repeat: the procedure is acceptable and fully possible only when the measurement conditions are fully known. In metrological practice, this circumstance may appear rather incidentally; thus, the approach should be allowed only in a very limited number of cases. However, assuming that the sizes of the applied antennas (probes) correspond to the limitations given by error δ_{1H}, with regard to inaccuracy of the error, it may be said that the accuracy of the magnetic field measurement is limited mainly by the magnitude of error δ_{2H}; i.e., the error results from the amplitude of measured field averaging along (on the surface of) the measuring antenna. At the same time, the most probable magnitude of the error corresponds to the assumption $\alpha = 1$. The identical conclusion has already been drawn at the end of Chapter 4.

In the end, it is worth mentioning the newly proposed method of magnetic field measurement based on the current in the antenna measurement using a current transducer. This method allows a remarkable increase in measurement sensitivity and stability; however, this is at the expense of frequency response uniformity [6]. The intensity of the current flowing in a loop antenna is directly proportional to the magnetic field strength, and the proportion is valid through a wide frequency range, which allows considerable simplification of a meter designed in accordance with this concept. Frequency response nonuniformity is affected mainly by the imperfection of the applied transducers. Let's express the hope that, after the development of such meters, this method will be dominant in the future—especially for higher frequencies.

Bibliography

[1] H. Whiteside, R. W. P. King. The loop antenna as a probe. *IEEE Trans.* vol. AP-7, No. 5/1964, pp. 291–294.

[2] R. W. P. King. The loop antenna for transmission and reception, Chapter 11 in T. E. Collin, F. J. Zucker, *Antenna theory part I*, McGraw-Hill 1969.

[3] P. Bienkowski. Accuracy limitation factors in near field EMF metrology, COST 281/EMF-NET, Seminar on the Role of Dosimetry in High-Quality EMF Risk Assessment; Zagreb, Croatia, 2006.

[4] J. D. Kraus. *Electromagnetics* (3rd edition). McGraw-Hill, 1984.

[5] S. Ramo, J. R. Whinnery. *Fields and waves in modern radio*. John Wiley & Sons, 1953.

[6] J. Zurawicki. *EM-field measurements* (in Polish). Graduate work at the Technical University of Wroclaw, Poland, 1992.

Chapter 6

Power Density Measurement

6.1 Power Density Measurement Methods

If electric, E, and magnetic, H, field strengths are known, the power density is explicitly determined by the Poynting vector **S**. The averaged value of the vector $\mathbf{S_a}$, which expresses the power flow from a source, is the subject of our interest. This quantity is given by:

$$\mathbf{S_a} = \frac{1}{2} \operatorname{Re} (\mathbf{E} \times \mathbf{H}^*)$$

(6.1)

In far-field conditions, it is enough to have (calculated or measured) one of the field components, as their mutual relationship is well known, and the relation is expressed by formulas in the form of Eq. (2.13) and Eq. (2.14). The power density is the sum of the electric, S_E, and magnetic, S_H, power densities, which may be expressed (changing vector notation to scalar) in the form:

$$S = S_E + S_H = 2S_E = 2S_H$$

(6.2)

In the near field, the mutual relationship of the E and H fields is unknown *a priori,* and it is a function of the structure of a radiation source as well as the distance between the source and the point of observation. Thus, a power density evaluation based on only one component measurement is loaded with error (method error). This error is of interest (and it creates a very important limitation in the use of the method and measuring equipment) because it is widely applied in commercially available measuring devices. However, acceptable conditions for equipment use are rarely discussed. We may add here that, although other methods of S measurement are known, and they are mentioned in other chapters of this work, the method discussed in this chapter is the sole technique used in wide-spectrum devices on the market.

Before we begin further considerations, we should focus our attention on two problems:

1. A method for direct power density measurement (similar to methods for electric and magnetic component measurements) remains unknown. This measurement requires that we find, as mentioned above, both field components. However, antennas that are sensitive to both components (with an exception discussed in Section 6.2) have resonant sizes and, in light of stated considerations, are useless for near-field metrology. It is known from the literature that power density measurement probe designs that fulfill our size requirements are composed of a number of E- and H-field sensors.

2. Without regard to applied measuring antennas and sensors types, for subsequent considerations, we will use only the field relations: the antenna and sensor parameters will not be taken into account. This assumption leads to the determination of the method error and, for the final estimation of measurement accuracy, it requires that the accuracy of the E- and/or H-field measurements are included as well. (These problems were considered in Chapters 4 and 5.)

Much as before, we here will accept assumptions that allow us to find the maximal values of errors that would majorize errors in measurements performed under other conditions. We will estimate errors for such sources as an elemental electric and magnetic dipole. As mentioned in Chapter 2, the EMF curvature around these sources is maximal (if we take into account sources of practical importance).

Expanding Eq. (6.1) and substituting Eq. (2.21), Eq. (2.22), and Eq. (2.23), for the monochromatic harmonic field we will have a time-averaged, complex magnitude of the Poynting vector:

$$
\mathbf{S}_m = \frac{1}{2}\left(E_R H_\phi^* \mathbf{1}_\theta + E_\theta H_\phi^* \mathbf{1}_R\right) = \frac{k\omega^3 \mu p^2}{32\pi^2 R^2}\left[1 + \frac{j}{(kR)^3}\right]\sin^2\theta \cdot \mathbf{1}_R
$$

$$
- \frac{jk\omega^3 \mu p^2}{16\pi^2 R^2}\left[\frac{1}{kR} + \frac{1}{(kR)^3}\right]\sin\theta\cos\theta \cdot \mathbf{1}_\theta
$$

$$(6.3)$$

where:

$\mathbf{1_R}$, $\mathbf{1_\theta}$ = versors,

   p = dipole moment [see Eq. (2.24)].

The power radiated by a source S_r is represented by the real part of Eq. (6.3). We shall note here that only this part of the vector can be transferred into heat in an absorbing body. This portion of the vector is expressed by a formula that is the same regardless of whether it is for a near or far field:

$$S_r = \frac{k\omega^3\mu p^2}{32\pi^2 R^2}$$

(6.4)

The magnitude given by Eq. (6.4) will be used in further considerations as the reference level for considered measurement methods.

6.1.1   Power Density Measurement by E or H Measurement

The validity of the power density measurement concept using E- or H-field measurement is doubtless in the far field. However, in the near field, one of the field components may be dominant and, as a result, the power density calculated on the grounds of the dominating component measurement will be overestimated, and *vice versa*. Taking components of the elemental electric dipole into consideration, it is easy to see that (for $\theta \to 0$) the following relationship is true:

$$\lim_{R\to 0} \left|\mathbf{E/H}\right| \to \infty$$

The subject of our measurement is the power density, which we will estimate based on the electric field measurement (electric power density) S_E, which for free space conditions and the source in the form of the electric elemental dipole, is:

$$S_E = \frac{\left|\mathbf{E}\right|^2}{2Z_0} = \frac{\omega k^3 p^2}{8\pi^2\varepsilon R^2}\left\{\frac{\sin^2\theta}{4}\left[1 - \frac{1}{(kR)^2} + \frac{1}{(kR)^4}\right] + \cos^2\theta\left[\frac{1}{(kR)^2} + \frac{1}{(kR)^4}\right]\right\}$$

(6.5)

Or, based on the magnetic field measurement (magnetic power density) S_H:

$$S_H = \frac{|\mathbf{H}|^2 Z_0}{2} = \frac{\omega k^3 p^2}{32\pi^2 \varepsilon R^2} \left[1 + \frac{1}{(kR)^2}\right] \sin^2\theta \tag{6.6}$$

The result of the measurement should reflect the total power density near a source and not the electric or the magnetic power density only. Thus, we emphasize here that, because of calibration conditions, contrary to Eq. (6.2), when Eqs. (6.5) and (6.6) were introduced, we had to assume:

$$S_r = S_E = S_H \tag{6.7}$$

Comparing the power densities given by Eqs. (6.4) through (6.6), we can define the error of the power density measurement near an elemental E-source by the electric field measurement η_{EE} and by the magnetic field measurement η_{EH}:

$$\eta_{EE} = \frac{|S_r - S_E|}{S_r} \tag{6.8}$$

and:

$$\eta_{EH} = \frac{|S_r - S_H|}{S_r} \tag{6.9}$$

Equations (6.8) and (6.9) are useful for measurement error estimations for limited error values of, say, 15 percent. For larger values of errors η_{EE} and η_{EH}, they become difficult to interpret and compare with other measurements. Meanwhile, the EMF strength and the power density measurements are among the least accurate measurements of physical quantities and often are accepted as satisfying the error value on the level ±3 dB, or even ±6 dB. Because of this, the errors plotted in Figs. 6.1 and 6.2, δ_{EE} and δ_{EH}, are defined as:

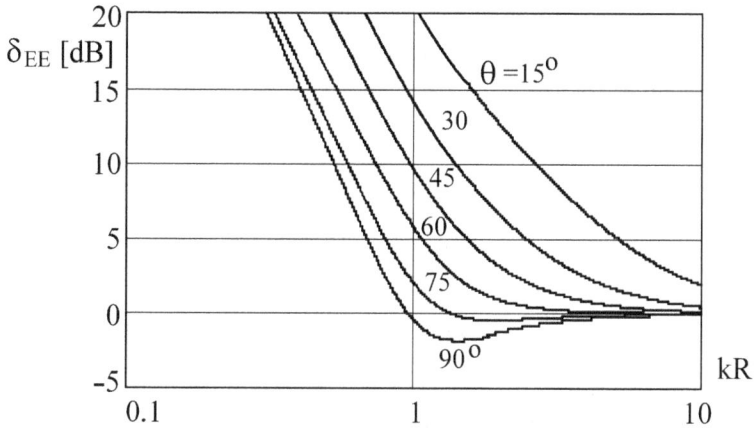

Figure 6.1 Error δ_{EE} as a function of kR.

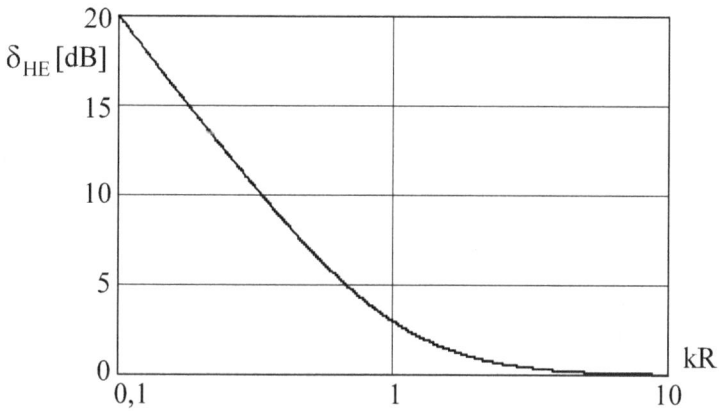

Figure 6.2 Error δ_{EH} as a function of kR.

$$\delta_{EE} = 10 \log \frac{|S_E|}{S_r} \tag{6.10}$$

$$\delta_{EH} = 10 \log \frac{|S_H|}{S_r} \tag{6.11}$$

Continuing similar considerations for a source in the form of an elemental magnetic dipole, we will have power density measurement errors in the proximity of the source δ_{HE} and δ_{HH} as a result of the use (for power density estimation around the dipole) of **E** and **H** measurements, respectively.

Instead of repeating calculations similar to the above, we will make the use of the analogy between E- and H-type sources (principle of symmetry), which allows us to write:

$$\delta_{EE} = \delta_{HH}$$
$$\delta_{EH} = \delta_{HE} \tag{6.12}$$

Presented considerations may be summarized as follows:

1. The power density measurement in the near field (especially in the proximity of electrically small sources) by the way of electric or magnetic field measurement is burdened with an error whose value depends on both the type and structure of the source and the measured EMF component, as well as the distance between the source and a point of observation.

2. For a source in the form of the elemental electric and magnetic dipole, the measurement error is always large for $\theta \to 0$, and it decreases for $\theta \to \pi/2$. While the E-field is measured near the magnetic dipole or H-field near the electric dipole, the error is independent of θ. It is possible to assume that, for $\theta \approx \pi/2$, the measurement errors do not exceed ± 6 dB for kR > 1.

3. Based on the curves shown in Figs. 6.1 and 6.2, as well as assuming minimal distances from a field source in which the measurements should be performed, it is possible to estimate the minimal frequency at which the measurement may be applied without necessity of using additional correction factors. (Because of the deterministic character of the error, the factors may be analytically estimated for a known source type, measured EMF component, propagation geometry, and distance).

4. Presented considerations make it possible to estimate the maximal error of the power density measurement in the neighborhood of the elemental sources at, for example, a distance of 5 cm. In both cases, the error exceeds 6 dB when measurements are performed at 300 MHz. Of course, the error value increases as frequency decreases (if the distance is kept constant).

The following two aspects of these considerations should be also emphasized:

1. The measurement method under consideration is widely used in power density meters available on the market; the meters "assure"

the power density measurements at frequencies even below 10 kHz without any explanation related to interpreting the results of such measurements. We should not be surprised by this approach by meter manufacturers, as requirements are spelled out in many standards (or standards proposals) that were (are) prepared with participation of the "best experts"!

2. These considerations make it possible to estimate maximal method errors. However, it has been shown [1] that measurements performed in proximity of physical sources (of finite dimensions) are not significantly lower—the statement is, in some sense, evident in light of Maxwell's equations. In close proximity to a source, the electric field is proportional to the charge of the source, while the magnetic field is proportional to its rate of change (current). This is a condition of a quasi-stationary field. In the stationary field, the E- and the H-fields may be considered to be independent of one other. An example of such a situation is the EMF around overhead power transmission lines, where the E-field appears while the line is connected to a voltage, but the H-field appears only if the line is loaded.

6.1.2   Power Density Measurement Using the Arithmetic Mean of S_E and S_H Measurement

This section addresses the subject of increasing measurement accuracy using a mean value approach to the electric and magnetic power densities.

As shown above, the results of power density measurement using electric field measurement near a source with the magnetic field dominating, and *vice versa,* are understated. At the same time, while the character of the source and the measured component are similar, the results are overstated. This might suggest that measurement of the arithmetical mean of the values might lead to an improvement in the accuracy of the power density measurement.

By analogy to the previous considerations, we will define the power density measurement error by way of the arithmetic mean of S_E and S_H measurement (η_a) in the form:

$$\eta_a = \frac{\left|2S_r - S_E - S_H\right|}{2S_r}$$

$$(6.13)$$

Because of the reasons presented above, the Eq. (6.13) will be modified to the logarithmic form (δ_a):

$$\delta_a = 10 \log \frac{\left| S_E + S_H \right|}{2S_r}$$

(6.14)

Calculated values of the error δ_a are plotted in Fig. 6.3.

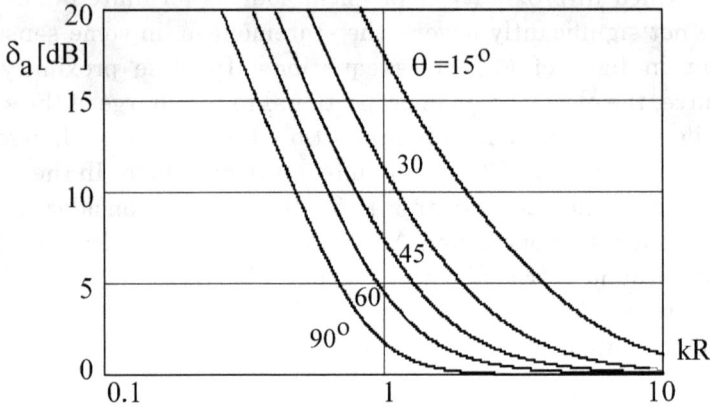

Figure 6.3 Error δ_a as a function of kR.

Because the principle of symmetry is valid here, Eq. (6.14) and the calculation results presented in Fig. 6.3 are identical for the case of an elemental electric dipole and a magnetic one. If we compare the curves shown in Figs. 6.1, 6.2, and 6.3, we can see that the error values in Fig. 6.3 are somewhat smaller than the previous ones. However, it is not a remarkable improvement, and the construction of a device for such measurement would be more complex and expensive.

Although it is not very clear why both power densities do not have "equality of rights" [2], it is why the power density is defined as a sum of 1/6 S_H and 5/6 S_E. However, in this case, an analysis of the accuracy improvement using a weighted arithmetic mean may be performed as well. The analysis was done, and it has shown that the measuring errors are very similar to those presented in the previous chapter.

6.1.3 Power Density Measurement As Geometric Mean of the S_E and S_H Measurement

Following the previously applied method, we will now estimate the power density measurement error (δ_g), near a source of radiation, using the measurement of the geometrical mean of the S_E and S_H. Similarly, as in the case of the arithmetical mean measurement, the results here are identical for both types of elemental sources.

We will define the measurement error directly in logarithmic form:

$$\delta_g = 10 \log \frac{\sqrt{\left| S_E S_H \right|}}{S_r}$$

(6.15)

The error calculation results are plotted in Fig. 6.4. In comparison to the curves shown in Fig. 6.3, a further leftward shift of the separate curves may be observed, which results in a further increase in measurement accuracy. However, it is not a remarkable improvement, and the above-formulated conclusion, relating to the complexity of necessary for the measurement equipment and its cost, remains valid.

For elemental sources, as well as for any open structure (with the exception of guided waves), we have **E** and **H** mutually orthogonal. Based on Eqs. (6.1) and (6.2), this may be written:

$$\left| S_m \right| = \sqrt{\left| S_E S_H \right|}$$

(6.16)

Thus, the curves shown in Fig. 6.4, apart from their role as discussed above, show the ratio of the power density modulus to the power density radiated by a source. Equation (6.16) is true for any arbitrary EMF source. Without regard to the unknown physical sense of the imaginary part of the power density, it would be possible to accept the geometric

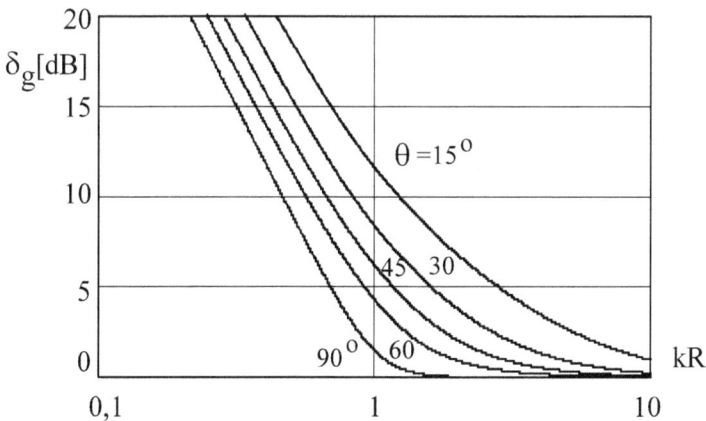

Figure 6.4 Error δ_g as a function of kR.

mean measurement results as maximizing any possible value of the power density; however, a protection standard based on such a concept would be too restrictive.

6.2 Power Density Measurement Using the Antenna Effect

The expression *antenna effect* is understood to refer to the susceptibility of a loop antenna to an electric field as well. As mentioned in Section 5.1 for near-field measurements, it is indispensable to use antennas of much smaller size than the minimal wavelength of the device's measuring band. They are sensitive only to **E** or only to **H**. This is only generally true if diameter of a circular loop antenna $D = 2r_0$ fulfills the following condition:

$$D \leq 0.01 \, \lambda_{min} \tag{6.17}$$

where λ_{min} = the shortest wavelength at which the antenna is to work.

If the condition of Eq. (6.17) is not fulfilled, the emf induced in the antenna by **E** is not negligibly small as compared to the emf induced in the antenna by **H**. As a result, the antenna is useless for these measurements. Equation (6.17) is related to a nonscreened and singly loaded loop, and it was introduced for plane waves.

A fascinating proposal of *antenna effect* application in EMF measurements was proposed by King [3]. The usefulness of the concept in different aspects of EMF measurements has been analyzed in detail [4]. Its practical use was proven in a device for near-field power density measurement. The essence of the concept is shown in Fig. 6.5.

A quadrant loop antenna of total circumference 1, made of a conductor of diameter 2a, is located on the xy plane of the Cartesian coordinate system (Fig. 6.5a). The resultant current induced in the loop is the sum of the magnetic field origin component, induced by magnetic field com-

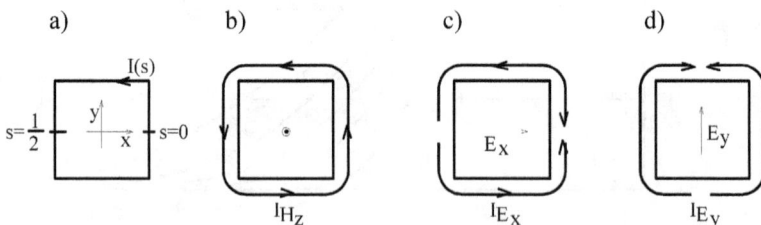

Figure 6.5 Currents in a doubly loaded loop antenna.

ponent H_z (Fig. 6.5b) orthogonal to the plane of the antenna, and that of the electric field, induced by the electric field components E_x and E_y (Figs. 6.5c and 6.5d, respectively). If we introduce two symmetrically located loads to the circumference of the antenna, for instance at s = 0 and s = 1/2, the currents at these points will be given by:

$$I(0) = I_{Hz} + I_{Ey} \qquad (6.18)$$

$$I(1/2) = I_{Hz} - I_{Ey} \qquad (6.19)$$

Equations (6.17) and (6.18) are valid when the current component I_{Ex}, associated with the E_x component, for s = 0 and for s = 1/2 is equal to zero. This is a basic assumption of the method, and the balance of the currents requires an appropriate antenna orientation in relation to the field components.

The operating principle of a power density meter, based on this concept, is shown in Fig. 6.6. Output voltages from two loads, symmetrically immersed in the loop winding, are fed to the inputs of two differential amplifiers. At their outputs, we obtain the voltage of the sum of the input voltages V_Σ as well as their difference, V_Δ:

$$V_\Sigma = 2 K_E E_y \qquad (6.20)$$

$$V_\Delta = 2 K_H H_z \qquad (6.21)$$

where K_E and K_H = transmittance for E and H field components.

Through a phase shift controller, these voltages are fed to a multiplier. The set is equipped with an output at which the voltage is proportional to the instantaneous value of the power density, and another one at which, as a result of the integrating circuit use, the voltage is proportional to the mean value of the power density. This measurement concept is based directly on the definition of the Poynting vector [Eq. (6.1)].

The presented construction is a certain expansion of solutions that for a long time were incorporated in radio direction finders working in Adcock and similar systems. The aim of research carried out in this

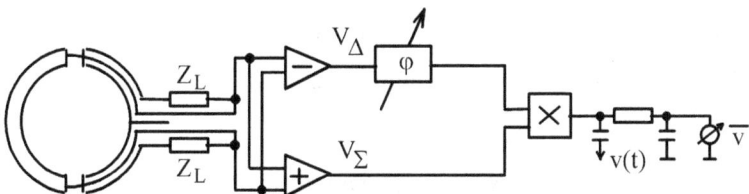

Figure 6.6 Block diagram of a power density meter with a doubly loaded loop

field at the Technical University of Wroclaw was to prove the possibility of constructing a meter for the real part of the Poynting vector measurement using a doubly loaded loop antenna, and to demonstrate that, although it is possible here to use an arbitrary complicated combination of dipoles and single loaded loops, the use of a doubly (multiply) loaded loop allows limitation of the antenna effect. In the case of a doubly loaded loop, Eq. (6.17) may be modified to the form [5]:

$$D \leq 0.15 \, \lambda_{min} \qquad (6.22)$$

Of course, multiplication of the loads leads to the possibility of further increasing the electrical sizes of a loop antenna without necessarily taking into account the size limitations caused by the presence of the antenna effect. However, these solutions and designs are useful only in the case of the far-field applications. This limitation is the result of averaging the measured magnetic field or power density at the surface of the antenna, which was discussed in Chapter 5.

The same concept was applied by Kanda in his construction of a power density meter [6]. In contrast to the design presented in Fig. 6.6, he used a photonic link for data transmission from both of the loop's loads to a measuring system. This solution allows an improvement in the insulation of the antenna against the meter and multiplies the possibilities of measured field interpretation as a result of preserved spectral and phase information.

The above-presented constructions allow us to formulate several comments:

1. The use of a doubly loaded loop antenna permits relatively simple power density measurement. At the same time, it is possible here to achieve quite high sensitivity as a result of the permissible use of larger-sized antennas [Eq. (6.17a)] as well as the introductory amplification of the sum and difference voltages.

2. The phase shifter applied in the meter, as well as transmittances of both channels, are a function of frequency (with regard to their phase and amplitude), and they considerably limit the applicability of the concept for discussed purposes, where wideband characteristic of the measurement is one of its dominant advantages.

3. The transmission of the signals from both loads to a meter may be troublesome because of deformation of the measured field by the leads. The use of a photonic link requires an increase in the probe's weight and/or complexity and, as a result, more expensive construction. However, photonic link technology will probably dominate in the future as a result of photonic technology advancements.

4. The design of the probe with the spherical directional pattern becomes more complex (see Chapter 7).

5. Equations (6.17) and (6.17a) were introduced for far-field (plane wave) conditions. In the near field, the probe's size limitation is more rigorous and impossible to use for determination *a priori* of an arbitrary source. This is especially of concern for electrically small electric-type sources for which the following inequality is fulfilled:

$$|\mathbf{E}|/|\mathbf{H}| \gg Z_0$$

To illustrate the statement, Fig. 6.7 presents the measured $Z_{E/H}/Z_0$ ratio for several types of Kathrain GSM900 type antennas, in a direction perpendicular to antennas' surface, as a function of R_0/D. $Z_{E/H}$ is defined as a ratio of measured values of E- and H-field strength. It may be called the *measured wave impedance* of the field at a point of concern, and D denotes maximal size of measured antenna. The figure shows that, in the direction of the main lobe, the ratio approaches unity for $R_0 > 10D$. However, if we accept a difference in the ratio at a level of about 10 percent, the distance $R_0 = D$ may be accepted.

6. The necessity of immersing the measuring antenna in the measured field so as to compensate for E_x is a factor that makes the measurement more difficult—especially measurements in the proximity of many non-correlated sources working at different frequencies, as well as under conditions of multipath propagation, not to mention their temporal variations.

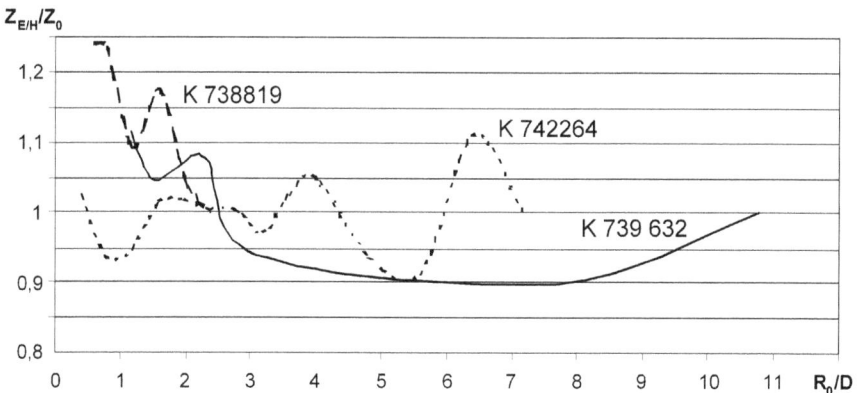

Figure 6.7   Measured $Z_{E/H}/Z_0$ as a function of R_0/D for three types of the Kathrain GSM900 antennas.

In order to illustrate problems relating to the structure of the described meter and its operation, we will consider the problem of frequency response matching of both the electric and magnetic parts of the device. Equation (5.4) presents a mutual relation of sensitivities when no means of frequency response shaping was applied. This problem was discussed in Chapter 4 in relation to electric field probes, and in Chapter 5 in relation to the magnetic ones. The formula implies the necessity of using such means so as to form a relationship of the sensitivities to both field components independent of the wavelength. The problem is much wider, however; it requires us to achieve appropriate transmittances for both EMF components. Because of the calibration conditions, with the use of the TEM wave, their mutual relationship should reflect the power relations in the plane wave. Apart from the necessity of having identical sensitivities for both components of the field, it is important to have identical shapes of their frequency responses as well.

In the most general case, the relation of transmittance (within the measuring band) of the electric field probe (creating a part of the power density meter) T_E to that of the magnetic field probe T_H may be defined in the form:

$$\frac{T_E}{T_H} = \frac{h_E \eta_E E}{h_H \eta_H H} \tag{6.23}$$

where:

h_E, h_H = effective lengths of the electric and the magnetic antennas applied,

η_E, η_H = attenuation factors of the detection circuitry of both the sensors.

Because of the calibration conditions, we must assume that $E = Z_0 H = 120\,\pi H$.

Substituting h_E in accordance with Eq. (4.23) and h_H as given by Eq. (5.8) for h, $r_0 \ll \lambda$, we will have:

$$\frac{T_E}{T_H} = \frac{\lambda h}{2\pi^2 r_0^2 n} \frac{\eta_E}{\eta_H} \tag{6.24}$$

The maximal sizes of the antennas are limited, as was discussed in Chapters 4 and 5 for electric and magnetic antennas, respectively. Based on the consideration of acceptable sizes for these antennas, and deriving from them permissible errors of both EMF components' measurement, we can write:

$$r_0 = a\,\lambda_{min}$$

$$h = ab\,\lambda_{min}$$

where:

a and b = constants, much smaller than unity,

λ_{min} = the shortest wavelength of the measuring band.

If we assume that $\eta_E = 1$ and $\eta_H = 1$ for $\lambda = \lambda_{max}$ or $\eta_H = \lambda/\lambda_{max}$ (where λ_{max} = the longest wavelength of the measuring band), then after substitution of these assumptions in Eq. (6.23), we will have:

$$\frac{T_E}{T_H} = \frac{b\lambda_{max}}{2\pi^2 an\lambda_{min}} \qquad (6.25)$$

Because $T_E/T_H \equiv 1$, thus:

$$\frac{\lambda_{max}}{\lambda_{min}} = \frac{2\pi^2 an}{b} \qquad (6.26)$$

Although the above considerations are considerably simplified, they allow us to draw the following conclusions:

1. In order to obtain equivalent sensitivities for both field components, we can accept that the applied antennas should fulfill the condition $h < r_0$.
2. The achievement of a wide range of measured frequencies is limited mainly by the small effective length of the magnetic antenna and by the relatively weak sensitivity of the magnetic field-measuring part of the device. It usually requires an artificial decrease in the sensitivity of its electrical portion.
3. The construction of the magnetic power density part of the meter creates problems despite the magnetic and electric antennas being of comparable sizes and the higher sensitivity and wider measuring band achievable for the electric field. Sensitivity and frequency response corrections are possible in both parts of the meter via

selection of appropriate geometrical sizes and shapes of the antennas as well as with the use of RC band-pass filters.

4. As the discussion shows, the condition $h < r_0$ is fulfilled in the majority of cases. Thus, it is possible to assume that the power density measurement error with an E/H probe (error of the measurement, not that of the method) is dominated by the error of the H-field measurement.

5. If $T_E \neq T_H$ in the frequency function or measured value function, the measuring error of the set increases. An example of the maximal value of the error is a case in which the probe is entirely insensitive to E- or H-fields and, as a result, the power density is then established on the basis of one component measurement only.

These issues are, perhaps, of excessively theoretical character. However, correct use of an arbitrary meter requires an understanding of its operating principles and, subsequently, limitations of its use and expected measurement accuracy under different conditions.

6.3 CONCLUSIONS AND COMMENTS

Widely applied methods of electromagnetic power density measurement were presented and analyzed in this chapter. The considerations have a purely theoretical character, and they are concerned with the field relationships only. The source of the measurement accuracy limitation, by way of E- or H-field measurement (method error), was demonstrated, and the magnitude of the error for different combinations of sources and measuring probes was estimated. Then, a certain accuracy improvement, as a result of simultaneous E- and H-field measurement, and a calculation of an arithmetic or geometric mean was proposed. The measurements of the mean values may be realized by two, simultaneously independent, measurements of these components using two different meters (one of them measures electric power density and the other magnetic). Then, the final result is obtained by computing the desired mean or by designing a probe (meter) that would be simultaneously sensitive to both components. The calculations could be done by a processor in the meter, and the ready result could be displayed. Construction of a meter that allows direct measurement of the real part of the Poynting vector was also outlined. Thus, there does exist the real possibility of measuring any aspect of the power density in the near field and in almost arbitrary conditions. Without regard to technical efforts to design appropriate equipment, increasingly precise meters available on the market and a better understanding the problem pose a question:

What is the sense of the power density and its measurement?

The quantity *power density* has primarily been introduced to the technique, and it is still successfully exploited, in antennas and propagation considerations—especially at microwave frequencies. A typical example of the quantity use is the *radiolocation equation* or widely applied term, *effective surface of an antenna.* It seems that the latter has had an especially strong influence on the impression of similarity between an antenna and the human body or other irradiated object. By way of simple multiplication of the power density by the effective surface of the absorber, the power (energy) absorbed by the object from an EMF should be obtained. In the case of object irradiation by a plane wave, such an approach could be, at least, accepted. In addition, estimation errors resulting from the measured field deformations caused by the object and the phenomenon of the energy dividing into its absorbed and reflected part, can be taken into account in the form of analytically estimated correction factors. These are dependent on the shapes and electrical parameters of the object, polarization of the field (in relation to the object), and other factors. In near-field conditions, such an approach, although theoretically imaginable, is nonsensical.

Analysis of the presented curves, illustrating the accuracy of currently-used power density measurement methods, permits the statement that at frequencies of below, say, 300 MHz, the measurement shows a large measuring error. Because of the deterministic character of the error, it is fully possible to estimate appropriate correction factors in every situation of practical importance, introduce them into results of a measurement, and in this way achieve extended measurement accuracy (if, that is, we do not take into account the possibility of using a meter that would enable a direct measurement of the real part of the power density, because of the cost of such a device, its complexity, or the lack of such devices on the market). As may be seen, it is possible to measure the quantity under almost any conditions. Apart from the aim of the presented considerations (i.e., EMF metrology designated for surveying and monitoring of EM environment and the resulting necessity of having the ability to perform quick and relatively simple measurement), we will return to the above formulated question from a bit wider perspective.

In order to answer the question, we shall refer to the EMF properties of the simplest elements in electrical engineering, i.e., a capacitor and an inductor. It should, of course be remembered that the elemental electric and magnetic dipoles, which were assumed here to be the basic sources in our considerations, are very similar to these elements from an electromagnetic aspect. The phase difference between the electric and magnetic fields inside these elements is equal to $\pi/2$ and, as a result, their vector product [expressed by Eq. (6.1)] is equal to zero. This means that, in this region, electromagnetic energy radiation does not appear or that electromagnetic energy losses do not exist in this area. The above does not mean that energy transfer in the area is altogether impossible. The best example here is a wide use of both elements for energy transfer, e.g., as heating electrodes in such applications as dielectric or inductive heating (Fig. 6.8). However, the energy transfer requires a lossy medium inside the capacitor or heating coil. We should add that the absorption model considered in Chapter 2 directly reflects such a situation while a lossy medium was immersed between the plates of a capacitor.

Correct power density measurement, i.e. without the measured field distortions (one of the most important requirements here and briefly outlined in Chapter 8), inside a capacitor or a solenoid must give a zeroth result. With regard to induction, we may widen the conclusion into the neighborhood of elemental (electrically small) sources and say that the measurement, in this aspect, makes no sense, because it does not reflect phenomena related to energy transfer. Completely reasonable, however, is the measurement of the temperature increase of a lossy dielectric (in the case of dielectric heating) or a lossy conductor or semiconductor (in the case of inductive heating), but this measurement depends on many other influences as mentioned in Chapter 2, and they radically limit the method's usefulness for discussed purposes.

The deliberations presented may be summarized as follows:

Power density measurement in the near field (in the aspect considered), regardless of the method applied, is senseless for frequencies below, say, 300 MHz. The same reservation should be applied to the application of power density, as a physical quantity, in any protection standard or measuring device devoted to measurement at frequencies below 300 MHz. There has been substantial improvement in this area of late, but even newly proposed standards are not free of weaknesses. To avoid misleading users, this quantity must not be applied in any

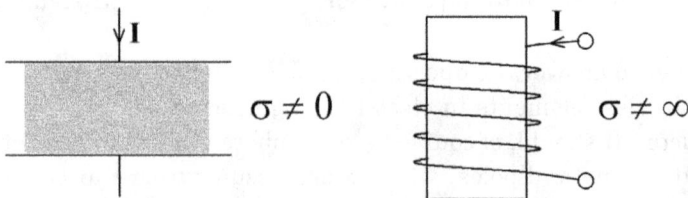

Figure 6.8 A lossy medium allows dielectric (left) or induction (right) heating.

aspect—not even for illustration or comparison purposes. On the other hand, the use of the quantity on microwaves, at frequencies above 300 MHz, is acceptable. But even here it is necessary to take precautions. We must recall once more an example presented in Chapter 2, in which the far field of a small dish antenna is similar to that of very tall long-wave antenna.

Of course, every physical quantity may be applied in any arbitrary way. The basic problem is here *understanding* its meaning. This does not change the above conclusion; on the contrary, it confirms that limitations related to the use of this quantity must be rigorous. People involved in the preparation of protection standards should note: if even the most experienced and competent persons working in this field have difficulty understanding the problems, then we can only speculate as to the challenges that face many common inspectors working in surveying-control services.

References

[1] H. Trzaska. Near field power density measurements. The First World Congress for Electricity and Magnetism in Biology and Medicine, Orlando, FL 1992. In: *Electricity and Magnetism in Biology and Medicine,* M. Blank (Editor), San Francisco Press, Inc., pp. 581–583.

[2] P. Pirotte. The problematic of ELF since 1970 and the actual situation with CENELEC pre-regulations. COST-244 WG Meeting, Athens 1995.

[3] H. Whiteside, R. W. P. King. The loop antenna as a probe. *IEEE Trans.* vol. AP-7, No. 5/1964, pp. 291–294.

[4] D. J. Bem, T. Wieckowski. On the measurement of hazardous EM field in lossy media using a small loop antenna. *Proc. 1981 Intl. EMC Symp.* Zurich, pp. 181–186.

[5] D. Hoff, G. Monich. Eine Sonde zur direkten Messung von Energiestrommungen im Nahfeld von Sende und Empfangsantennen (in German). *NTZ,* No. 27/1974 Heft. 8, pp. 313–318.

[6] M. Kanda. An electromagnetic near-field sensor for simultaneous electric and magnetic field measurement. *IEEE Trans.* vol. EMC-26, No. 3/1974, pp. 102–110.

Chapter 7

Directional Pattern Synthesis

When radio communication or propagation problems are discussed, an idealized understanding of the expression *linear* or *circular* (or, more accurately, *elliptical*) *polarization* is usually used. At the same time, in order to simplify any complicated problems, it is "forgotten" that even a linearly polarized wave propagating near the surface of a lossy medium changes its polarization to an elliptical one that is applied to measure the equivalent conductivity of the medium. In the case of multi-path propagation of a circularly (elliptically) polarized wave as a result of multi-path interference, a spatial rotation of the polarization plane may be observed. Thus, it may be called *quasi-spherical* or *quasi-spheroidal polarization*. Usually, such a situation occurs in the neighborhood of a complicated system of radiators (primary and secondary) when they are excited with an FM modulated signal or, for instance, as a result of changes in phase differences of the rays coming to a point of observation, due to frequency changes or Doppler effect.

Under these conditions, three spatial components of the measured EMF vectors may appear at an observation point; this was taken into account for the estimation of measuring errors due to spherical pattern changes of an electric field probe (Chapter 4) and of a magnetic one (Chapter 5). The above, however, in no way changes the observation (Section 3.1) that, at any arbitrarily selected moment in time and space, there exists one and only one electric field vector and one magnetic field vector. The expression *polarization* implies the possibility of drawing in space, by the head of a vector, complex curves (elliptical polarization) or planes (ellipsoidal polarization). The complexity shows that metrological difficulties may occur here that can lead to problems with measurement result interpretation and therefore a decrease in measurement accuracy.

The considerations presented in Section 4.4 lead to the following conclusions:

- An arbitrary radiating system, composed of elements sensitive to an electric or magnetic field, if it is small in relation to wavelength, has a sinusoidal directional pattern. (A specific case of *antenna effect* use, as an example of simultaneous sensitivity to both field components, was presented in Section 6.2.)

- It is impossible to construct an antenna that is insensitive to EMF polarization (an omnidirectional one) using any number of arbitrarily oriented elements in space while they are connected to a common load.

- In order to obtain a spherical (omnidirectional) pattern, it is necessary to use at least three mutually orthogonal antennas, and their output voltages should be summed after squaring them.

- It is worth remembering that herein we discuss in detail the properties of omnidirectional E-field probes. However, the designs of magnetic probes are almost identical, as are the directional patterns of electrically small electric and magnetic radiators (sensors).

7.1 A PROBE COMPOSED OF LINEARLY DEPENDENT ELEMENTS

Let's consider a probe, designated only for elliptically polarized field measurements, composed of "n" identical dipoles intersecting in the center of the xy plane of a Cartesian coordinate system. The angles between them are identical and equal π/n, and the first dipole crosses the axis y at angle γ as shown in Fig. 7.1. The probe is illuminated by a monochromatic plane wave of angular frequency Ω propagating in the z direction. Its electrical field components are given by:

$$\mathbf{E} = \mathbf{1}_x A \sin\Omega t + \mathbf{1}_y B \cos\Omega t \tag{7.1}$$

where A and B are amplitudes.

If we assume that the sizes of the system are small in comparison to the wavelength, the influence of the mutual couplings between these antennas is negligible, and if all of the antennas are loaded with square-law detectors, then, based on the above conclusions, it is possible to describe their directional patterns sinusoidal. Thus, the square of the output voltage of the i-th antenna (v_i) is:

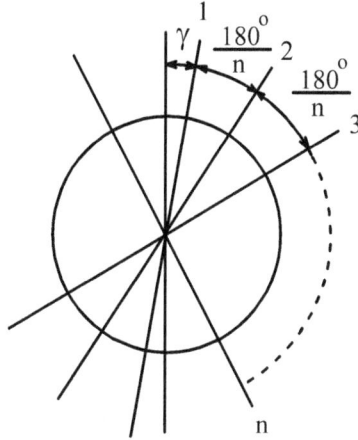

Figure 7.1 System of n dipoles.

$$v_i^2 = \frac{h^2}{T} \int_0^T \left[A \sin \Omega t \cos (\mathbf{1}_x, \mathbf{1}_i) + B \cos \Omega t \cos (\mathbf{1}_y, \mathbf{1}_i) \right]^2 d(\Omega t) =$$

$$= \frac{h^2}{2} \left[A^2 \cos^2 (\mathbf{1}_x, \mathbf{1}_i) + B^2 \sin^2 (\mathbf{1}_x, \mathbf{1}_i) \right]$$

$$(7.2)$$

where $\mathbf{1}_i$ = versor of i-th dipole.

In order to calculate the output voltage from n dipoles, we will make use of the generalized Pythagorean theorem, introduced specially for this purpose. It may be formulated in the following form:

If, in a circle, are inscribed n secants that intersect at the center of the circle, the angle between two adjoining secants is π/n, and the first of them crosses the axis of symmetry of the circle at angle γ, then the sum of squares of the directional sines (and cosines) of the secants is constant and equals n/2.

We may write this in the form:

$$\sum_{i=1}^n \cos^2 \left(\frac{\pi}{n} i + \gamma \right) = n/2$$

$$\sum_{i=1}^n \sin^2 \left(\frac{\pi}{n} i + \gamma \right) = n/2$$

$$(7.3)$$

Equations (7.3) are valid for $n \geq 2$. The terms used in the formulas are identical to those in Fig. 7.1. We may notice that when n = 2, the formulas are equivalent to the traditionally understood Pythagorean theorem.

If take a sum of Eq. (7.2) making use of Eqs. (7.3) for n detectors, we will have the square of the output voltage V of the system.

For elliptical polarization, it is:

$$V^2 = \sum_{i=1}^{n} v_i^2 = \frac{nh^2}{4}\left(A^2 + B^2\right)$$
(7.4)

For circular polarization, A = B and:

$$V = Ah\sqrt{\frac{n}{2}}$$
(7.5)

For linear polarization, for instance, B = 0 and:

$$V = Ah\,\frac{\sqrt{n}}{2}$$
(7.6)

The above allows us to formulate the following two conclusions:

1. The results [Eqs. (7.5) and (7.6)] reveal the evident variation of the probe output voltage depending on the measured field polarization. However, regardless of that, the known voltage V does not make it possible to find *a priori* the magnitudes of A and B and then the polarization of the field.
2. The output voltage is a function of n. Thus, when probes containing electric and magnetic antennas are designed for power density measurement, it is indispensable to assume identical values of n for both EMF components.

In the presented estimations, it was assumed that h $\ll \lambda$. This has made it possible to simplify the calculations. The approach, however, leads to an error resulting from radiation pattern changes when the ratio of h/λ increases. The problem was discussed in Section 4.4, and it may be assumed that the factor can be neglected with accuracy of δ_{1E}.

The directional properties of the system were discussed for the probe illumination by the plane wave when the probe is designated for near-field measurements. Changes of the spherical pattern as a function R_0/h were discussed in detail in Section 4.4, and it may be assumed that the considerations are valid in the case of any combination of antennas considered here.

Equations (7.3) are of purely trigonometric nature and their use for the pattern calculations is limited to not very large values of n, when mutual couplings between individual dipoles are negligibly small.

This discussion was presented to illustrate problems considered here from one side, but they also are accurate enough for practical applications from the other.

7.2 SPHERICAL RADIATION PATTERN OF AN E/H PROBE

Examples of applied versions of E/H sensors are shown in Fig. 7.2. In the solutions shown in Figs. 7.2a and 7.2c, the roles of the electric field and magnetic sensors are separated, and there appears to be the possibility of independently shaping their frequency responses in accordance with the considerations presented in Section 6.2. The solution shown in Fig. 7.3, like those presented in Fig. 6.6, allows full functional separation of both sensors. The simplest solution, shown in Fig. 7.2b, does not include such a possibility. As a result, it may be used for power density measurement only at a singular, selected frequency at which the calibration of the probe was performed. Its main advantage is the simplicity of its construction.

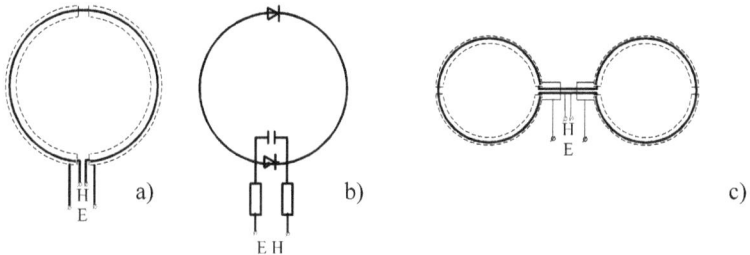

Figure 7.2 Examples of E/H probes: (a) set with divided electrostatic screen, (b) probe with doubly loaded loop, and (c) probe with two loop antennas.

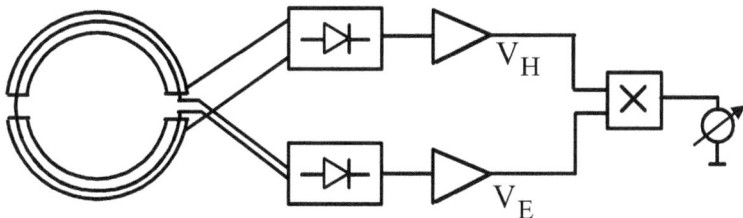

Figure 7.3 Block diagram of the power density meter.

All of the designs shown in Fig. 7.2 suffer from the common disadvantages of very strong coupling between antennas of both EMF components as well as ineffective performance of the electrostatic screen divided into two parts (Fig. 7.2a). Whereas the solution using two-loop antennas (Fig. 7.2c) is too large, especially for near-field measurements, where the effective length of the loops (even multiturn ones) is small in comparison to that of the electric dipole.

In Section 7.1, we analyzed the possibility of constructing an EMF sensor for elliptically polarized field measurement using a combination of linearly dependent antennas, applied to available probe designs to increase their sensitivity (the output voltage is then proportional to the square root of the number of antennas applied = n). However, practical solutions require a slightly modified approach, i.e., the use of n quadrant antennas with one common arm. The common element is placed perpendicularly to the plane created by n unipoles. The detection diodes are connected between the common unipole and their respective unipoles at the plane.

The electromotive force e at the output of a short quadrant antenna, as shown in Fig. 7.4, under similar illumination conditions as assumed in considerations presented in section 7.1, when it is illuminated from the direction of the bisector of the angle β, is:

$$e = C \cos \left(\gamma + \frac{\beta}{2} \right) \sin \frac{\beta}{2} \tag{7.7}$$

where C = a constant, and other indications are as in Fig. 7.4.

If we assume in Eq. (7.7) that $\gamma = 0$ and $\beta = \pi/2$ (which reflects the above-presented construction of the probe), we will obtain a twofold reduction of the emf induced in such quadrant antenna with relation to a symmetrical dipole of identical sizes. Thus, the benefit of the solution is evident, especially if we take into account that the usual number of

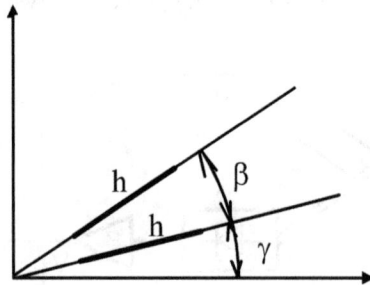

Figure 7.4 The quadrant antenna.

unipoles, n > 20. The directional pattern of an individual quadrant antenna is (in accordance with the above) sinusoidal, and the direction of its maximum is identical to that of the bisecting line of the antenna angle β. Thus, pairs of the quadrant antennas, which create a common plane, are sensitive to the circular polarization in the plane. Hence, if n = 4, or a multiple of it, and any unipole is positioned symmetrically in relation to the others, it makes it possible to synthesize a spherical pattern of such a system of quadrants.

A similar concept has already been applied in the construction of an E/H sensor [1]. If we use only one loop antenna immersed inside of an electrostatic screen (e.g., as a half of the set shown in Fig. 7.2c), and place three such loops on three separate planes of the Cartesian coordinate system in such a way that their inputs are located close to the center of the coordinate system, the system allows us to synthesize the spherical pattern for the magnetic field. However, the use of the electrostatic screens of these loops as arms of three quadrant antennas does not assure the spherical radiation pattern for the electric field. If we suppose that the shape of the radiation pattern for the magnetic field is satisfactorily uniform, then the screens of the loops can be replaced by unipoles of equivalent effective length h. To satisfy the requirement of a spherical pattern for the electric field, it would be necessary to add an auxiliary unipole to the system, located symmetrically in relation the to the previous ones as shown in Fig. 7.5.

The system of unipoles shown in Fig. 7.5 contains six quadrant antennas that have common arms. Three of the quadrants are created successively by pairs of unipoles 1–3, while the next three quadrants are created by the unipole 4 and unipoles 1–3, respectively. It can be assumed that the lengths of the unipoles 1–3 are identical and that

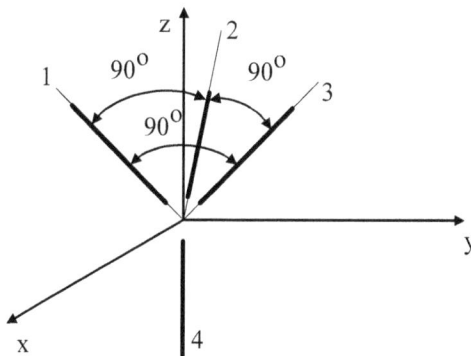

Figure 7.5 Spatial unipole orientation of a modified triple-quadrant antenna.

they are infinitely thin. The analysis shows that the length of the uni-pole 4 (h_4) should be:

$$h_4 = h \left(2 - \sqrt{3}\right) = 0.268h$$

However, unipoles 1–3 cannot be assumed to be infinitesimally thin because of their roles in the electrostatic screens of the loop antennas. Hence, in order to obtain the spherical pattern of the electric field probe, it is necessary to enlarge unipole 4 by a factor of two or more and, depending on the screen sizes, we will have:

$$h_4 = h \, (0.5 \div 0.75)$$

The considerations outlined above are approximate and incomplete. They were employed here to illustrate the methods and concepts of spherical pattern synthesis. They were necessary, however, to aid an understanding and evaluation of the technical solutions available on the market. They (sometimes) look very attractive, but their construc-tion and parameters may exclude their applicability to many measure-ments, especially those performed in the near field.

7.3 A PROBE COMPOSED OF THREE MUTUALLY PERPENDICULAR DIPOLES

In Section 4.4, it was shown that, to obtain an omnidirectional pattern, it is necessary, for example, to use three mutually perpendicular anten-nas whose output voltages should be summed after squaring.

Let's consider a system that fulfills these conditions:

Three identical, mutually perpendicular dipoles (a, b, and c) are located in the Cartesian coordinate system as shown in Fig. 4.11 (p. 66). The system is illuminated by a monochromatic plane wave polarized ellipsoidally. Its electric component may be expressed, for example, in the form [2]:

$$\mathbf{E} = \mathbf{1}_x \, A \sin \Omega t + \mathbf{1}_y \, B \cos \Omega t \cos \omega t + \mathbf{1}_z \, C \cos \Omega t \sin \omega t \tag{7.8}$$

where:

    A, B, C = constants (amplitudes),

         ω = angular frequency of the polarization ellipse rotations, let $\Omega \gg \omega$,

$\mathbf{1}_x, \mathbf{1}_y, \mathbf{1}_z$ = versors of the coordinate system.

The emf e_i induced by the field in i-th dipole can be expressed as:

$$e_i = hA \sin \Omega t \cos (1_x, 1_i) + hB \cos \Omega t \cos \omega t \cos (1_y, 1_i)$$
$$+ hC \cos \Omega t \sin \omega t \cos (1_z, 1_i) \tag{7.9}$$

where 1_i (i.e., 1_a, 1_b, and 1_c) = versors of respective dipoles located in the Cartesian coordinate system "a," which is arbitrarily rotated in relation to the system shown in Fig. 4.11.

The procedure of the emf e_i squaring may be performed twofold. We will discuss both possibilities.

7.3.1   The Vector Summation

We will define the vector summation procedure based upon the summation of the output emf of the three dipoles e_i (i.e., e_a, e_b, and e_c corresponding to the dipoles a, b, and c, respectively) after squaring them. The summation is realized with phase information conservation and without any additional filtration procedures that would limit the measurement frequency band.

When quasi-ellipsoidal polarization is considered, the output voltage is equal to the sum of the squared three emfs given by Eq. (7.9), and it is expressed by complex formulas that are difficult to interpret. In order to simplify the considerations (and the final formulas), we will assume that the measured field is polarized quasi-spherically, i.e., $A = B = C$. Thus, the sum of the emf squares, after some simple modifications, may be expressed as:

$$V^2 = h^2 A^2 \left\{ \sin^2 \Omega t \left[\cos^2(1_x, 1_a) + \cos^2(1_x, 1_b) + \cos^2(1_x, 1_c) \right] \right.$$
$$+ \cos^2 \Omega t \cos^2 \omega t \left[\cos^2(1_y, 1_a) + \cos^2(1_y, 1_b) + \cos^2(1_y, 1_c) \right]$$
$$+ \cos^2 \Omega t \sin^2 \omega t \left[\cos^2(1_z, 1_a) + \cos^2(1_z, 1_b) + \cos^2(1_z, 1_c) \right]$$
$$+ 2\sin \Omega t \cos \Omega t \cos \omega t \left[\cos (1_x, 1_a) \cos (1_y, 1_a) \right.$$
$$+ \cos (1_x, 1_b) \cos (1_y, 1_b) + \cos (1_x, 1_c) \cos (1_y, 1_c) \right]$$
$$+ 2\sin \Omega t \cos \Omega t \sin \omega t \left[\cos (1_x, 1_a) \cos (1_z, 1_a) \right.$$
$$+ \cos (1_x, 1_b) \cos (1_z, 1_b) + \cos (1_x, 1_c) \cos (1_z, 1_c) \right]$$
$$+ 2\cos^2 \Omega t \sin \omega t \cos \omega t \left[\cos (1_y, 1_a) \cos (1_z, 1_a) \right.$$
$$+ \cos (1_y, 1_b) \cos (1_z, 1_b) + \cos (1_y, 1_c) \cos (1_z, 1_c) \right] \right\} \tag{7.10}$$

It is easy to remark that the expressions in the first three brackets represent sums of the squares of directional cosines of two rectangular coordinate systems rotated in relation one to the other, and the sums are equal to unity. The expressions in the next three brackets represent products of respective terms of two rows or two columns of matrices of directional cosines of one coordinate system in relation to the other one, and they are, by definition, equal to zero. After some transformations, we have:

$$V = h\,A \tag{7.11}$$

For a circular polarization, there is no rotation of the polarization plane ($\omega = 0$) and, after some transformations, similar to the above and reduced to the plane of polarization, we will have a result identical to that given by Eq. (7.11). For the linear polarization, there is only one spatial component of E-field, and E is given by Eq. (7.8) in which, for example, $B = C = 0$. Taking this into account in the above considerations, we have for the linear polarization:

$$V = h\,A \sin \Omega t \tag{7.11a}$$

Depending on the detector type used in the sensing system, for the linear polarization, we will have a result identical to that obtained for quasi-spherical and circular polarization using peak value detection. It differs from it by 3 dB when mean square detection is applied.

7.3.2   A DC Summation

For direct current summation, we will define the procedure based upon the respective dipoles loading with the RMS detectors and a summation of their output voltages after filtration of the alternating components. Such a procedure is widely applied in the construction of omnidirectional EMF probes designated for measurements related to human safety and protection of the natural EM environment as well as within the entire EMC area.

We will calculate the square of the output voltage of i-th detector using the value of emf e_i given by Eq. (7.9):

$$
\begin{aligned}
V_i^2 &= \frac{h^2}{T\tau} \int_0^T \int_0^\tau e_i^2 \, d(\Omega t)\, d(\omega t) \\[2mm]
&= \frac{h^2}{4}\left[A^2 \cos^2(1_x,1_i) + \frac{B^2}{2}\cos^2(1_y,1_i) + \frac{C^2}{2}\cos^2(1_z,1_i) \right]
\end{aligned}
\tag{7.12}
$$

where T and τ = periods equivalent to angular frequencies Ω and ω, respectively.

After summation of the output voltages V_i of the three dipoles, we have:

• For the quasi-ellipsoidal polarization:

$$V = \frac{h}{2} \sqrt{A^2 + \frac{B^2 + C^2}{2}}$$
(7.13)

• For the quasi-spherical polarization:

$$V = \frac{h\,A}{\sqrt{2}}$$
(7.14)

• For the elliptical polarization:

$$V = \frac{h}{2} \sqrt{A^2 + B^2}$$
(7.15)

• And for the linear polarization:

$$V = \frac{h\,A}{2}$$
(7.16)

We should note here that similar results lead to a procedure of RMS detection of the output voltages when vector summation is used [Eqs. (7.11) and (7.11a)]. The difference between both the summations (in a constant, nonessential from the presented considerations point of view) is a result of the assumption of the half-period detection.

7.3.3 Achromatic Field Measurement

Until now, we have considered directional patterns of several probe systems illuminated with a harmonic field. Very often, it may be necessary to perform EMF measurements in conditions where M spectral components appear simultaneously. We will define the m-th electric field component in the form:

$$E_m = 1_{xm} A_m \sin \Omega_m t + 1_{ym} B_m \cos \Omega_m t \cos \omega_m t + 1_{zm} C_m \cos \Omega_m t \sin \omega_m t$$
(7.17)

Equation (7.17) is identical in shape to Eq. (7.8), the only exception being the index "m," which differentiates the m-th spectral component. The proposed notation does not emphasize directions, amplitudes, or frequencies of M spectral components of the E-field that appear in a specific point in space (the point where an EMF probe is placed—the probe is composed of three mutually perpendicular dipoles). The other assumptions here are identical to those accepted before. However, in order to obtain the required results of estimations, an additional assumption is indispensable here, i.e., the assumption of the probe's flat frequency response. This means that the effective length h is a frequency-independent, constant value. The requirement is easy to fulfill when, in the measured spectrum, only spectral components appear in the probe's medium-frequency band (as discussed in Chapters 4 and 5). When the condition is not fulfilled, in the Eq. (7.17), it is necessary to introduce multipliers that would normalize the transmittance of the probe to that in its medium-frequency range. The resulting field, illuminating the measuring device, is equal to the sum of its M components (frequency fringes). These components are of arbitrary frequencies, without the requirement of any harmonic relation between them that could suggest phase synchronization or similar phenomena. As noted previously, the spatial location of any component may be an arbitrary one that should not affect the results of calculations (measurements). Then the calculation procedure is identical to that described in Section 7.3.2.

The electromotive force e_i induced by EMF in the i-th dipole is:

$$e_i = h \left\{ \sum_{m=1}^{M} \left[A_m \sin \Omega_m t \cos(1_{xm}, 1_i) + B_m \cos \Omega_m t \cos \omega_m t \cos(1_{ym}, 1_i) + \right. \right.$$
$$\left. \left. + C_m \cos \Omega_m t \sin \omega_m t \cos(1_{zm}, 1_i) \right] \right\} \quad (7.18)$$

Making use of this formula, we will calculate the square of the output voltage of i-th detector:

$$V_i^2 = \frac{h^2}{T_m \tau_m} \int_0^{T_m} \int_0^{\tau_m} e_i^2 \, d(\Omega_m t) \, d(\omega_m t) \quad (7.19)$$

Exchanging the order of the summation and the integration, we may write:

$$V_i^2 = \frac{h^2}{4} \sum_{m=1}^{M} \left[A_m^2 \cos^2(1_{xm}, 1_i) + \frac{B_m^2}{2} \cos^2(1_{ym}, 1_i) + \frac{C_m^2}{2} \cos^2(1_{zm}, 1_i) \right] \quad (7.20)$$

Now we will reduce the Eq. (7.20) for different polarizations:

- When the M signals are polarized quasi-ellipsoidally, the probe output voltage is:

$$V = \frac{h}{2} \sqrt{\sum_{m=1}^{M} \left[A_m^2 + \frac{B_m^2 + C_m^2}{2} \right]}$$

(7.21)

- For M signals polarized quasi-spherically:

$$V = \frac{h}{\sqrt{2}} \sqrt{\sum_{m=1}^{M} A_m^2}$$

(7.22)

- For M signals polarized elliptically:

$$V = \frac{h}{2} \sqrt{\sum_{m=1}^{M} \left(A_m^2 + B_m^2 \right)}$$

(7.23)

- And for M signals polarized linearly:

$$V = \frac{h}{2} \sqrt{\sum_{m=1}^{M} A_m^2}$$

(7.24)

Also in this case, the results obtained with the use of the vector summation and RMS detection are similar, which again confirms the equivalence of both procedures.

A more important conclusion is that the voltages given by Eqs. (7.21) through (7.24) are the measure of the effective (RMS) strength of the field containing more than one spectral fringe. Thus, it is not a sum of the effective magnitudes but the effective magnitude of the sum (resultant field).

7.4 COMMENTS AND CONCLUSIONS

Detailed consideration of polarization problems is needed for omnidirectional probe design and construction and, primarily, to understand and properly apply them. These needs flow from, among other issues:

- The specificity of the polarization intricacies, especially in the near field.
- The necessity of understanding the polarization phenomena in order to select optimal procedures for proper measurement planning and interpretation of the results.
- The development of an ability to select an appropriate probe and meter for specific measurement conditions.
- The achievement of s skills needed for crucial evaluation of commercially available devices.

The directional properties of a probe have a fundamental importance for its use in variously polarized EMF measurements. Below, we will briefly summarize the application ranges as well as advantages and disadvantages of probes with selected directional properties. Directional properties of different probe types are illustrated in Fig. 7.6.

A. Linear EMF Polarization, Sinusoidal Probe Pattern

This combination is especially characteristic of EMF measurements in the far field. In the near field, one may ask, "How is it known that the polarization is linear?" This reveals an advantage of a probe with a sinusoidal pattern: it makes it possible to check whether the EMF polarization is linear and, if it is, it permits localization of the spatial orientation of the **E** (or **H**) vector. The probe's directional properties allow us to localize an unknown radiation source. In the case of linear polarization, the maximal indications of a meter appear when the measuring dipole is placed parallel to the **E** vector [Eq. (4.37)]. On the other hand, when the dipole is placed in a plane perpendicular to the **E** vector, the indications should be minimal; the latter is valid for an arbitrary position (rotation) of the dipole on the plane.

B. Elliptical EMF Polarization, Sinusoidal Probe Pattern

A good example of an EMF with stable elliptical polarization in time and space is the field under overhead HV transmission lines. The polarization plane here is perpendicular to the lines. Measurement using a sinusoidal probe allows us to find parameters of the polarization ellipse and, thus, the maximal value of the **E** (and **H**) field strength. However, measurement of the effective (RMS) value of the field requires two measurements and then calculation of the result of the measurement as a square root of the sum of the squares of the two results. Both measurements should be performed on the polarization plane in two, mutually perpendicular positions of the probe. If there is any doubt regarding the field polarization, it is sufficient to check whether the indications of the meter vanish in the direction perpendicular to the polarization plane, i.e., parallel to the wires of the line.

C. Ellipsoidal (Spherical) EMF Polarization, Sinusoidal Probe

The term *ellipsoidal (spherical) polarization,* as already mentioned, has no physical sense. However, it is a convenient description of a situation in which the polarization ellipse rotates in space with a frequency (which may be a function of time and space), as discussed in detail in Section 7.3. Investigated EMF here has three, non-zero, spatial components. It leads to non-zero indications of a meter while the measuring probe is arbitrarily oriented in space. The advantage of using the sinusoidal probe here is, again, the possibility of measuring the spatial properties of the investigated field and finding the maximal strength (as to its magnitude and direction) of the **E** (and **H**) vector. But a determination of the effective value (RMS) involves the same procedure as above, with its expansion to the third dimension.

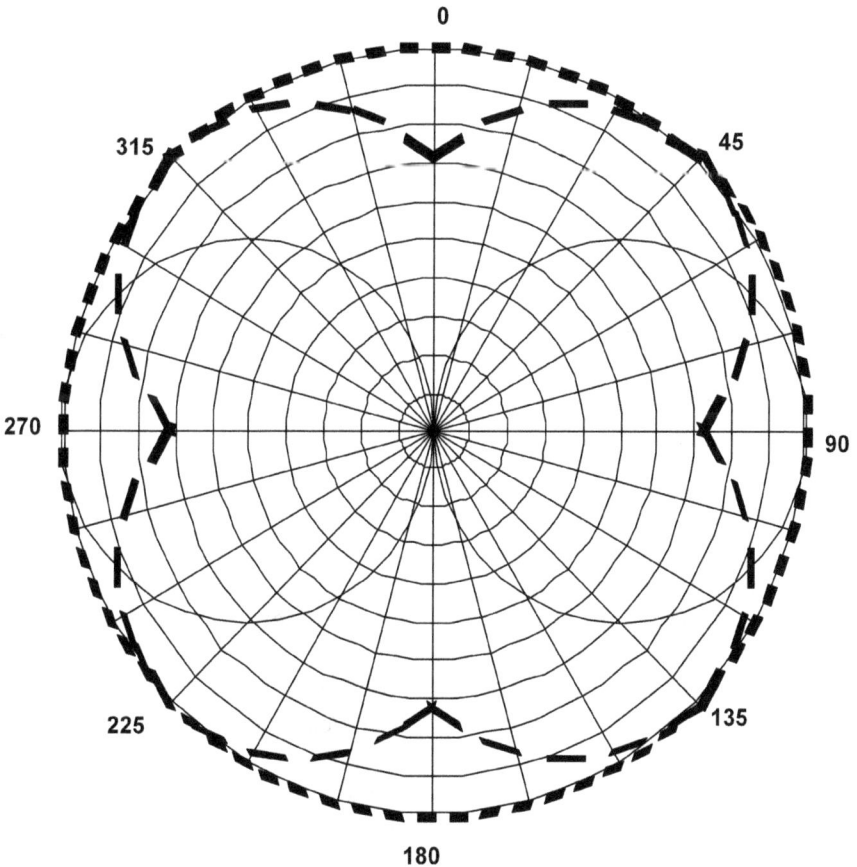

Figure 7.6 Directional patterns of different types of probes: sinusoidal (continuous line), directional for linear voltage summarization (dotted line), and omnidirectional for squared voltage summarization (dashed line).

D. Linear or Elliptical Polarization, Circular Probe Pattern

The probe with a circular pattern still has not been mentioned. The simplest example of this probe consists of two mutually perpendicular dipoles loaded with square-law detectors. While using this probe for measurements of linearly or elliptically polarized fields, it is sufficient to immerse the probe in the field and orient it in the space so as to produce maximal indications on the meter. It is certainly possible to use such a probe for polarization evaluation, i.e., to prove whether the polarization is linear or elliptical (minimal meter indications, while the **E** vector of the investigated field is perpendicular to the plane at which the measuring antenna of the probe is located, suggest linear polarization). However, it is impossible to absolutely determine whether the polarization is elliptical or quasi-ellipsoidal, and no simple procedure exists that would allow the probe to be used in conditions of polarization unknown *a priori*. This feature of the probe eliminates its practical applicability for our purposes; however, there is one exception: measurements in the proximity of overhead HV transmission lines, where the polarization is well known.

E. Arbitrary Polarization, Spherical Probe

This measurement consists only of introducing a probe into the measured field and reading indications on the meter. In surveying or monitoring services, this procedure has advantages. It is necessary to remember, however, that the probe does not permit us to discover any information on the polarization properties of the investigated field, to measure maximal intensity of a spatial field component, or to detect the direction to an unknown field source.

F. Unknown Probe

With a number of manufacturers offering probes, it may be useful to confirm their directional properties ourselves. The simplest way is to use a linearly polarized field generated, for instance, between two plates (E-field) or the simplest circular coil (H-field). Using the procedures presented above, it is possible to determine the directional properties of a probe. For example, a sinusoidal probe should have only one maximum, a probe with circular pattern should show no indication of change when it is rotated such that the E-vector lies in the polarization plane of the probe, and an omnidirectional probe should be insensitive (within limits indicated in its manual) to its arbitrary rotations in the field. The test should be performed at low frequency if possible.

As may be seen from the considerations presented herein, an optimal measuring probe design does not exist. Meters have long been available that allow us to measure one, two, or three spatial components of the field, which is equivalent to including all of above-discussed probe types. However, because of their complexity and the associated risk of making mistakes during measurements, as well as their high price tags, they have not gained widespread use until recently.

Ignoring polarization problems and thus making an inappropriate measuring probe choice may lead to measurement errors well exceeding 50 percent. This, even taking into account the low "resolution" of bioeffects and generally low accuracy of field measurements, may be unacceptable. We should remember that all considerations regarding directional properties were performed for plane-wave illumination conditions. The above-mentioned measuring error (50 percent) was also estimated for conditions when the measuring probe has enough separation that only polarization could be dominant for the estimation. When measurements are performed in proximity to primary or secondary radiation sources, the error may be much larger due to source–probe mutual couplings (as discussed in Chapters 4 and 5) and other factors.

References

[1] T. Babij, H. Trzaska. Power density measurements in the near field. IMPE 1976 Microwave Power Symp., Leuven, Belgium. Publ.: *Journ. of Microwave Power* No. 2/1976, pp. 197–198.

[2] E. Grudzinski, H. Trzaska. Polarization problems in the near field EMF measurements. 36 Internationales Wissenschaftlisches Kolloquium Ilmenau 1991.

Chapter 8

Other Factors Limiting Measurement Accuracy

In previous chapters, we discussed the most typical and most frequently appearing factors limiting the accuracy of EMF measurements. They resulted mainly from the type of antenna applied in an EMF probe—particularly its size, and mutual couplings between the antenna and a radiation source. In this chapter, we will briefly discuss, among others topics, the influence of thermal drift on the probe parameters, the role of its dynamic characteristics, deformation of the measured field by a person performing the measurements and by the meter, resonant phenomena, and others. At the end of the chapter, an example based on the uncertainty budget estimation of EMF measurement will be presented.

8.1 THERMAL STABILITY OF A FIELD METER

Because the parameters of semiconducting devices, such as diodes, magnetodiodes, Hall cell, etc., used in EMF probes are strongly thermally dependent, the meter's functional parameters are also a function of temperature. Thermal instability may be a source of large measurement errors. The errors (such as the deterministic ones) may be limited considerably by the use of appropriate correction factors (calculated or measured) or via the choice of the measurements conditions that assure minimal errors. Although the builders of the measuring devices do their best to obtain the most positive parameters (including thermal) in their devices, more and more extremely simplified devices are appearing on the market. These devices can assure a comparatively low price, but sometimes this comes at the expense of quality. In order to enable a user to evaluate these devices under extreme conditions, the role of quality must be examined. This is discussed below.

The subject of our analysis will be a probe for an electric field measurement, containing a dipole antenna loaded with a diode detector. This is the simplest version of the probe and the most susceptible to the influence of temperature changes. However, similar probes are widely used for electric field and power density measurements. The susceptibility of magnetic field probes for variations in ambient temperature is similar to that of an electric field probe, as their designs are similar. Although any element applied in an arbitrary type of measuring probe is more or less susceptible to temperature changes, the most sensitive is the detection diode. In order to improve thermal stability, a thermocouple instead of a diode is sometimes used. It has the important advantage of the identical shape of the dynamic characteristics over the full measuring range, which ensures true RMS indications, but its range is very limited, its sensitivity is insufficient, and it may be easily damaged when overloaded. The diode detectors are often initially polarized so bridge diode detectors and other approaches are applied in order to stabilize their working point and temperature changes. As mentioned above, we would like to illustrate the maximal values of the errors (and not just thermal errors), which is why the simplest solution was chosen for this analysis.

From the considerations presented in Chapter 4, we know that the parameters of the detector (diode) may affect transmittance within the measuring frequency range [Eq. (4.14)], and as a result, the sensitivity of the meter, the shape of the transmittance in the lower frequency range [Eq. (4.13)], and the lower corner frequency [Eqs. (4.15) and (4.15a)]. The importance of these effects will be outlined consecutively.

8.1.1   Sensitivity Variations within the Measuring Band

Variations in meter sensitivity due to temperature change can be defined as follows:

$$\delta_1 = \frac{T_0 - T_0'}{T_0}$$

(8.1)

where:

T_0 = transmittance of the probe within the measuring band in the temperature of the probe calibration,

T_0' = transmittance of the probe in the ambient temperature where the measurements are performed.

The transmittance T_0, in the light of the Eq. (4.14), is a function of the detector input capacitance. The capacitance consists of the parasitic capacitance of the montage, capacitance of the filter, and the diode capacitances: the diffusion capacitance C_D and the junction capacitance C_J. For instance, the capacitance C_D is given approximately by [1]:

$$C_D = \frac{q}{kT}(I_0 + I_s)\frac{\tau}{2} = g_d\frac{\tau}{2}$$

(8.2)

where:

q = the electron charge,

k = Boltzmann's constant,

T = temperature,

τ = carrier's relaxation time,

I_0 = DC component of the diode current,

g_d = dynamic conductance of the diode,

I_s = saturation current, while:

$$I_s \approx I_{s0}\exp(b\,\Delta T)$$

(8.3)

I_{s0} = saturation current in the introductory temperature,

ΔT = temperature rise,

b = constant ($b \approx 0.07$/K for germanium diodes and 0.1/K for silicon ones).

An element of the temperature influence is also capacitance C_J and the dynamic characteristics of the diode. While the dependence of the specific parameters of the diode as a function of temperature is complex, it also depends on the selected working point of the diode and its loading resistance. Usually, in diode detectors and especially in diode rectifiers, it is assumed that $I_s \ll I_0$, which allows simplification of Eq. (8.2) and indicates a decrease in the temperature effects on the diode parameters. In our case (which results from the diode loading with a transparent, high-resistivity transmission line), the simplification is not acceptable, especially in the most sensitive measuring ranges. As a result, we encounter problems with stabilization of the detector's thermal conditions, and there is a variation in the probe parameters when the temperature changes.

Returning to the analysis of thermal stability and sensitivity, Eq. (4.14) correctly defines transmittance in a "high-frequency" sense. Here, the temperature influences are seen throughout variations in capacitances C_D and C_J, or more precisely, throughout changes of these capacitances in relation to the other capacitances of the probe. By way of an artificial extension of the detector's capacitance, it is possible to improve the thermal stability in this aspect: it will be followed by a reduction in the lower corner frequency. However, this will happen at the expense of the sensitivity reduction. Here, however, there is a DC aspect of the problem as well. Because the detector's work with low current, the result is high loading resistance (as mentioned above). Considering the detector as the DC source of conductance given by Eq. (8.2), and taking into account Eq. (8.3), based on Thevenin's theorem, it is easy to see that the DC part of the probe is not free of thermal instabilities.

A systematic approach to the problem is not presented here because of its complex character and the dependence of the phenomenon on many factors, including the selected configuration of the probe and applied protective means. Detailed discussions are available in the literature [2, 3, 4]. In order to summarize our considerations and comments and to illustrate the scale of the problem, Fig. 9.1 shows correction curves applied in the EMF meter type NFM-1, manufactured in the former German Democratic Republic and still used in many Eastern European countries [5]. (The problem is considered in various publications, e.g., Ref. [6].) Respective curves, prepared (experimentally) for different magnitudes of measured fields, give the correction

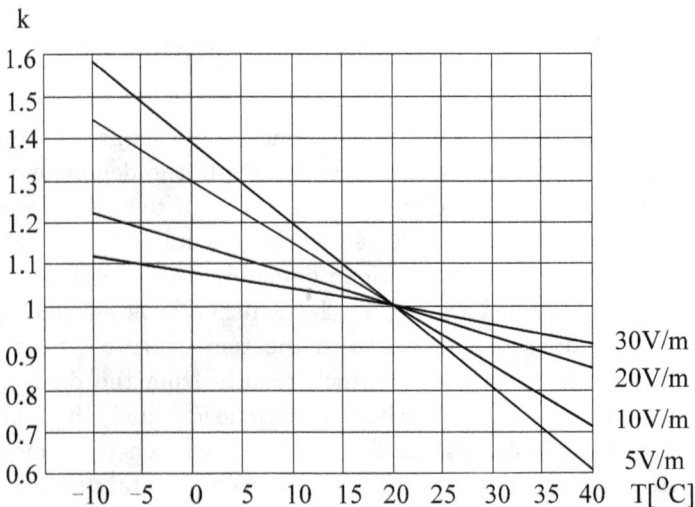

Figure 8.1 Correction coefficients of the NFM-1 type field meter.

factors that are to be multiplied by the value indicated on the meter to obtain the measurement result. We should notice here the increase in the meter's sensitivity with the temperature rise. The maximal magnitudes of the factors are as much as 2 percent/K while field intensity on the level of 5 V/m is measured, and the influence of temperature on sensitivity vanishes for larger intensities of measured fields and the transition of the detector's working point to the linear part of its dynamic characteristics. Contrary to applied practice, the curves should be corrected periodically because of variations induced by the aging process; moreover, the curves were prepared *en masse* for the meters, whereas thermal stability is a function not only of the type of the detection diode but also on the diode itself.

In order to illustrate the problem of diode temperature sensitivity, several measurements were performed. The results of measurements for an electric field probe with a germanium diode, and similarly with a Schottky barrier diode, are presented in Fig. 8.2. A temperature coefficient δ_t is formulated as follows:

$$\delta_t = 20 \lg\left(\frac{V_{Tmax}}{V_{Tmin}}\right) \text{ [dB]}$$

(8.4)

where:

V_{Tmax} = a voltage lead to the diode in temperature 50°C,

V_{Tmin} = the voltage in temperature 5°C.

The temperature coefficient shows a difference in voltages that should be led to a diode in different temperatures so as to have identical detected output voltage.

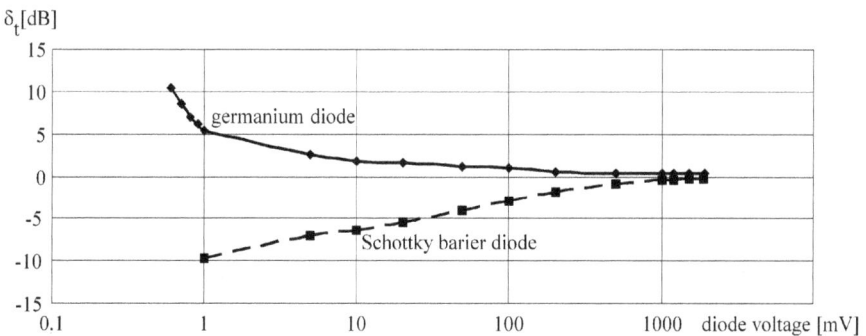

Figure 8.2 Measured temperature coefficient δ_t for a probe with a germanium diode (solid line) and with a Schottky barrier diode (dashed line).

A conclusion may be drawn from presented measurement results: the temperature dependence vanishes while output voltages of the diode are in the linear part of its characteristics; i.e., above about 0.2 V for germanium diodes and about 0.7 V for silicon ones. Unfortunately, in the most sensitive ranges of an EMF meter, the dependency is significant, as may be deduced from Fig. 8.1.

8.1.2   Shape Changes of the Frequency Response and the Lower Corner Frequency

With the aim of determining the temperature dependence of probe sensitivity not only as a function of the detector capacitance but also as a function of its conductance, it was necessary to cite Eqs. (8.2) and (8.3). The dependence of the lower corner frequency on both of these parameters directly results from formulas defining the frequency [Eqs. (4.15) and (4.15a)]. Apart from the above comments, related to the components of the detector capacitance, we should note that, in these formulas, R_d represents parallel connection of the dynamic resistance of the diode and the loading resistance, i.e.:

$$\frac{1}{R_d} = g_d + \frac{1}{R_t + R_i}$$

(8.5)

where:

$R_t =$ resistance of the transparent line,

$R_i =$ input resistance of the line load (a DC voltmeter).

The resistance, as may be seen from Eq. (4.13), is of fundamental importance for the shape of the frequency response in the low-frequency range.

In general, changes in the frequency response outside of the measuring band are as important as those within the band, both for the accuracy of the measurement and the sensitivity (transmittance) within the band. Let's formulate several comments:

- In a thermally noncompensated probe (the one considered here), the thermal changes in the lower corner frequency and the shape of the frequency response (sensitivity) within the low-frequency range are of the same order of magnitude as the above-shown changes in the sensitivity. However,

- The thermal sensitivity compensation within the measuring band is not enough to obtain this compensation outside the band.

- From this we can draw the conclusion that there is a hidden source of possible measuring faults when a probe is used (making use of the known shape of its frequency response outside the measuring band) for measurements outside the band. It is sometimes suggested in meter manuals (as, for instance, in the case of the aforementioned NFM-1 meter) in relation to frequencies below the band.

- Experience shows that all the thermal effects depend on the type of the diode applied and may even be different for different specimens of the same type. Further effects may be observed due to aging and various external factors. For example, overheating the diode while assembling the probe or as a result of overloading it affects both the dynamic characteristics of the detector and the thermal parameters of the diode; after overheating, the parameters of the diode drift over a relatively long period of time and never return to their initial state. By the way, this suggests the necessity of applying an ongoing check of the meter's parameters (probe, sensor).

- It is important to notice that both the shape of the frequency response and the lower corner frequency are functions of the magnitude of measured field strength; the dependence is especially noticeable if the changes of the lower corner frequency are compared to the field rate δ_{fl} as defined by:

$$\delta_{fl} = \frac{\Delta f_l}{f_l} \frac{1}{\Delta E}$$

(8.6)

where Δf_l and ΔE are changes in the lower corner frequency and the rate of the measured E-field, respectively.

The measured magnitudes of δ_{fl} are [7]:

- for the detector with a germanium diode: (0.01 to 1) percent m/V,
- for a detector with a Schottky barrier diode: (0.05 to 2) percent m/V.

The effects presented here are a function of the measured value; they are the strongest in the square-law part of the characteristics and vanish after the detector transitions to the linear part of the characteristics. The phenomenon is illustrated in Fig. 8.3 [8] (reproduced with permission of IEEE).

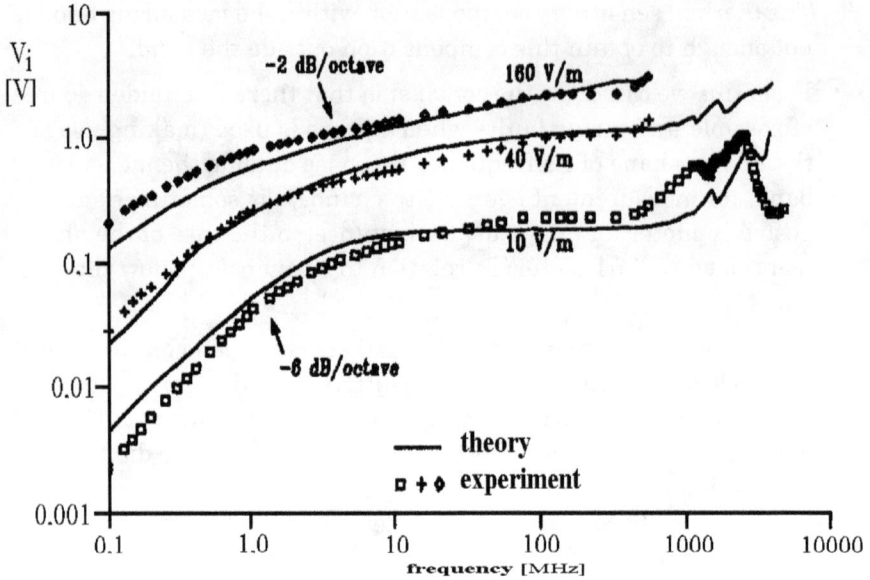

Figure 8.3   Frequency response alternations as a function of measured E-field [8].

8.2 THE DYNAMIC CHARACTERISTICS OF THE DETECTOR

The static characteristics of an ideal semiconducting diode in the range of permissible currents (i) and voltages (V_d) is given by:

$$i = I_s \left[\exp \left(\frac{q}{kT} V_d \right) - 1 \right]$$

(8.7)

The analytical calculation, based on Eq. (8.7), of the DC voltage magnitude at the output of the probe, i.e., the voltage at the input of a DC voltmeter represented by its input resistance R_i (Fig. 4.3), requires the solution of a nonlinear differential equation of the first order. We will cite here results of considerations presented by M. Kanda, performed for a monochromatic exciting signal and stabilized thermal conditions, based on the equation solution by P. F. Wacker [9]. We will complete the solution with terms representing the use of the frequency response shaping filter and dividing the detector output voltage on the divider created by the resistance of the transparent line and the input resistance of the voltmeter.

For small values of V_d, the DC voltage at the voltmeter input V_i is:

$$V_i = - \frac{q}{4kT} \left(\frac{e_A \, C_A}{C_A + C_d + C_f} \right)^2 \frac{R_i}{R_i + R_t}$$

(8.8)

Whereas for large values of the voltage V_d:

$$V_i = - \frac{e_A \, C_A}{C_A + C_d + C_f} \frac{R_i}{R_i + R_t}$$

(8.9)

Where terms in the formulas are as shown in Fig. 4.3.

As may be seen from the above, the dynamic characteristics of the probe consists of three segments. These are (1) low values of V_d [for which the characteristics are of square-law character, as given by Eq. (8.8)], (2) high voltages V_d (where it is linear), and (3) medium voltages V_d (where the shape of the characteristics is transitional from square-law to linear). The measured characteristics $V_i = f(E)$ for a typical specimen of an AE-1 probe are shown in Fig. 8.4.

The probe sensitivity variations due to thermal changes in its dynamic characteristics were outlined in Section 8.1.1. In order to emphasize the importance of these temperature variations, Eq. (8.7) was cited. However, apart from the probe's susceptibility to temperature variations, presented in Figs. 8.1 and 8.2, the dynamic characteristics of the probe (detector) shows its change from the square law to the linear one (if we omit a relatively narrow transitional segment). The shape of the characteristics is completely nonessential when monochromatic harmonic fields are measured. But the phenomenon is of primary importance when measurements are performed in complex achromatic

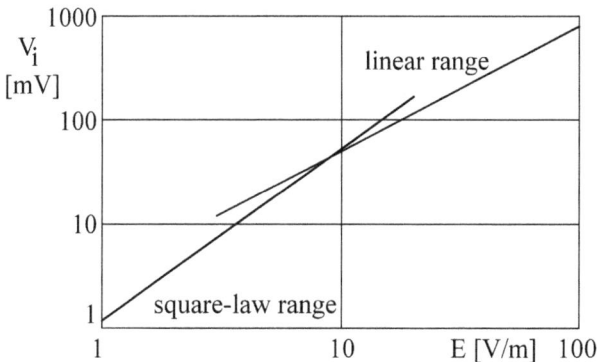

Figure 8.4 Dynamic characteristics of an AE-1 probe measured for T = 21°C.

fields and in pulsed fields. We will illustrate this with several simple examples.

A. Two or More Harmonic Fields at Different Frequencies

In Section 8.3.3, it was shown that if we measure cw EMF, containing m spectral harmonic components, with the use of a square-law detector, we will have a result proportional to the RMS value of the measured field. This was the theory. Let's compare it with the practice.

In the case of measuring two or more EMF harmonics (E_n) and a square-law detection, the effective RMS value (E_{RMS}) of the resultant field intensity should be equal to the square root of the sum of squared intensities of separate components:

$$E_{RMS} = \sqrt{\sum_n E_n^2} \, a$$

(8.10)

It was necessary to first assume that the applied detector is of square-law dynamic characteristics. In the case of the linear part of the characteristics, a meter will show a value proportional to the sum of the amplitudes of the EMF components whereas, in the case of the transitional part of the (diode) detector characteristics, the indication will be somewhere between the sum of the amplitudes and an RMS value given by Eq. (8.10).

Now let's formulate an RMS value measurement error of a resultant field δ_{RMS} that allows a comparison of measured EMF intensity, read from meter indications (E_{in}) and the expected value, calculated on the grounds of Eq. (8.10):

$$\delta_{RMS} = 20 \lg \left(\frac{E_{in}}{E_{RMS}} \right) \, [dB]$$

(8.11)

A series of measuragments were performed, with the use of different EMF probes, at different frequencies and with a different number of EMF components. In order to illustrate the considerations, below we present the results of measurements with the use of three probes and at two frequencies, i.e., 900 and 1800 MHz. For this case, Eq. (8.11) may be presented in the form:

$$\delta_{RMS} = 20 \lg \left(\frac{E_{in}}{\sqrt{E_{900Mhz}^2 + E_{1800Mhz}^2}} \right) \, [dB]$$

(8.12)

Results of the measurements are shown in Fig. 8.5.

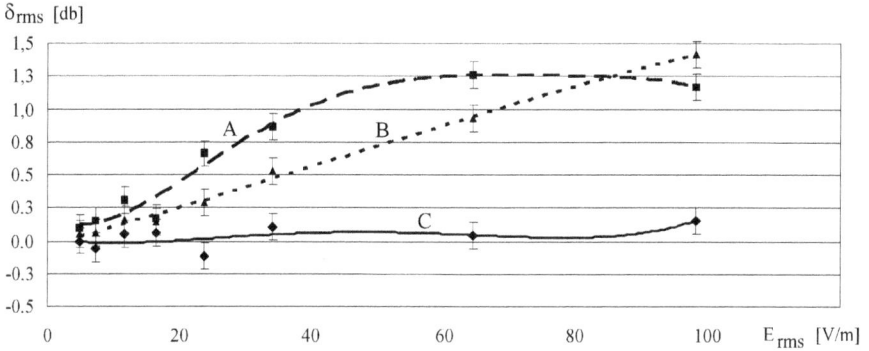

Figure 8.5   RMS value measurement error in achromatic EMF with the use of three EMF probes, labeled A, B, and C.

During the measurements, cw E-fields at frequencies of 900 and 1800 MHz were generated and measured separately and then simultaneously. The error δ_{RMS} was estimated using Eq. (8.12). A similar experiment may be simply repeated by any user of a meter (probe), and no standard EMF is necessary for a meter (probe) parameter evaluation of aspects of its dynamic characteristics.

B. AM Modulated Field, Square-Law Detection

Above it was shown that, even while cw EMFs are measured, errors may appear due to shape of the dynamic characteristics of the applied measuring device. Now we will consider two cases of EMFs modulated in amplitude.

The amplitude modulated voltage V_m may be show in the form:

$$V_m = A (1 + m \cos \omega t) \cos \Omega t \qquad (8.13)$$

where:

  $A =$ amplitude,

  $m =$ modulation depth,

  $\omega, \Omega =$ angular frequencies of the modulating signal and the carrier wave, respectively.

If the signal is detected with a square-law detector, we will have its effective voltage V_{ieff} given by:

$$V_{ieff} = \frac{A}{\sqrt{2}} \sqrt{1 + \frac{m^2}{2}} \qquad (8.14)$$

The first term of the formula represents the effective value of the harmonic carrier wave, while the rooted term shows the role of the modulation depth. The term varies from unity (for m = 0) to about 1.25 (for full modulation, i.e., for m = 1).

C. AM Modulated Field, Peak Value Detection

If we assume that in Eq. (8.13) magnitudes of both cosines equal unity, we will obtain the amplitude of the DC voltage at the output of the linear detector V_{ia}:

$$V_{ia} = A(1 + m) \tag{8.15}$$

If we assume m = 1, we will see that the amplitude of the voltage equals 2A.

Figure 8.6 presents comparative results of measurements of three different EMF probes in a 100 percent AM modulated field as a function of the field intensity. The figure allows the conclusion that meter C really has a square-law detector, while meter A is equipped with a diode detector that causes a transition in the meter's response from the square-law in lower measuring ranges to the linear one for higher field intensities. The dynamic characteristics of meter B suggest a diode detector with a short time constant that limits its ability to indicate an increase, due to modulation, in higher-intensity ranges. Figure 8.6 and Eq. (8.14) suggest a simple method of checking characteristics. If the tested meter is immersed in quite a strong field (right side of Fig. 8.6) and we regulate depth of modulation from m = 0 to m = 100 percent, meter indications should not increase by more than about one fourth if the characteristics are square-law, and the

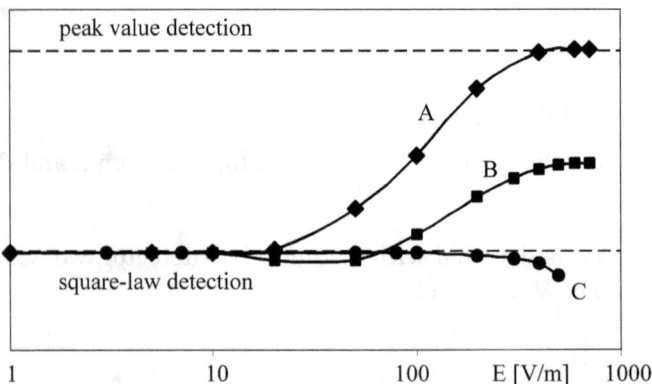

Figure 8.6   Three EMF probes in 100 percent AM EMF.

indications may increase even twofold while the characteristics are linear.

D. Detection of a Manipulated CW Signal

Let's repeat the above considerations for a periodic signal of a duration time T in which the carrier wave appears during period τ. Consequently, we will have:

- For the square-law detection:

$$V_{ieff} = A \sqrt{\frac{\tau}{T}}$$

(8.16)

- And for the linear detection:

$$V_{ia} = A$$

(8.17)

Theoretical estimations and experiments show that, in this case, apart from the above-discussed dependence of the meter's response to the dynamic characteristics of the detector, an important role is played by the time constant of the applied meter (probe). We may notice here that the phenomenon may play a primary role while nonstationary EMFs are measured; for instance, fields generated by rotating radar antennas where trains of pulses are measured at the point of observation.

In order to illustrate a role of the pulse duration time τ in relation to the period T, Fig. 8.7 presents results of estimations of the pulsed EMF measurement accuracy defined as:

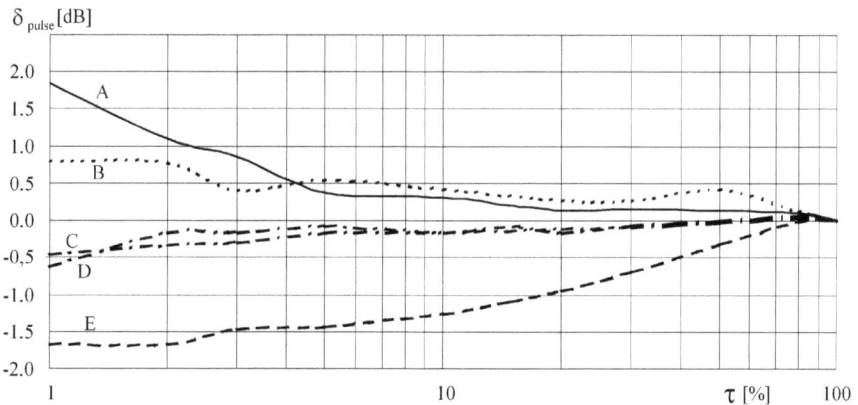

Figure 8.7 Pulse-modulated field measurement error of selected types of commercial EMF probes, labeled A through E.

$$\delta_{pulse} = 20 \lg \left(\frac{E_{in}}{E_{RMS}} \right) \text{ [dB]}$$

$$(8.18)$$

where:

E_{in} = indications of a meter,

E_{RMS} = RMS value of measured EMF.

The measurements were performed with the use of several commercially available EMF probes at a frequency of 900 MHz and pulse repetition frequency of 217 Hz (which corresponds to the GSM slot repetition time) as a function of τ/T in percent. The figures do not allow us to differentiate a source of the measuring error (dynamic characteristics of the detector or the meter's time constant); however, it shows that measurements, especially of short pulses, may be loaded with large errors—again, not to mention nonstationary EMF measurements.

Each of the presented estimations and measurements provides correct indication results for a diode detector-equipped probe. For our purposes, both the effective and peak values are of concern in aspects under consideration. The problem is in the necessity of precisely reducing the probe's measuring range to limits within which the dynamic characteristics are known. For economic reasons, the problem is often neglected. As a result, quite large measuring errors may appear when measurements of nonmonochromatic fields, especially pulsed ones, are performed. If, in the case of the AM modulation, we may assume that a twofold difference between RMS and peak value measurement is acceptable, in the case of pulsed fields, the measurement error may reach an arbitrary value. These errors may be successfully limited by way of precise demarcation of the linear and the square-law segments of the dynamic characteristics. This procedure can be done for any type of meter via analysis of the character of the meter's indicator scale or its calibration curves (if applicable).

8.3 MEASURED FIELD DEFORMATIONS

When a measurement is performed, the measuring probe must be immersed in the measured field. The probe is usually accompanied by the meter (indicating device) and a measurement technician. Any material medium placed in a homogeneous field causes field deformation. Depending on the electrical size and electrical properties of the medium, the deformation may mean a change in the spatial distribution of the equipotential lines, reflection, or diffraction. The latter two phenomena usually appear when the sizes of the medium are comparable or larger in comparison to the wavelength, while the former is specific to (quasi-) static fields [10]. The problem must be well known and understood by people performing the measurements on the one hand, and it must be mentioned in methodology textbooks for them and in appropriate standards on the other. As an example, we will use Polish standards and publications of the authors involved [11, 12]. In order to illustrate the importance of the problem, we present two case studies:

1. Figure 8.8 presents results of estimations completed for a meter immersed in a homogeneous field while the meter is placed parallel to the E-field force lines (right) and perpendicular to them (left) [13]. The situation is specific for low-frequency fields.

2. The influence of the measurement technician on measurement results were analyzed and experimentally verified. A typical *in situ* measurement is presented in Fig. 8.9a. If we neglect reflections from the floor and walls, the propagation between a source and a probe consists of two rays, i.e., a direct ray and a ray reflected from the technician. To estimate the significance of this phenomenon, experimental studies in laboratory conditions were performed using the setup shown in Fig. 8.9b. In the experiment, a metallic plate of dimensions 1.5×0.7 m represented the person. The distance between probe and the plate was changed from 5 cm to 3 m.

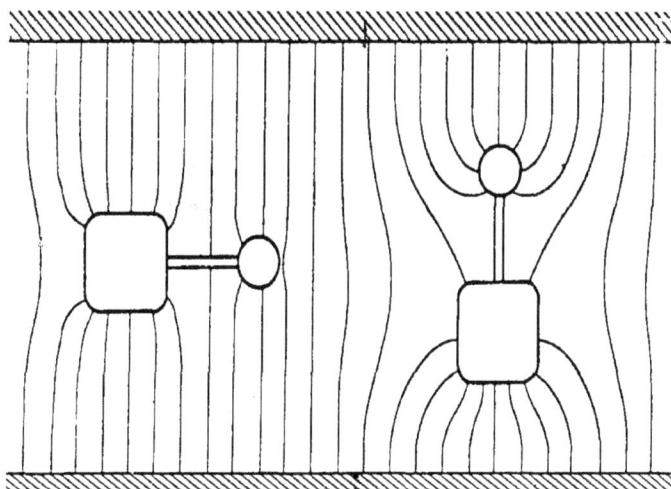

Figure 8.8   EMF deformations caused by a measuring device [13].

(a)

Trx Antenna

D

EMF meter

Metal plate

(b)

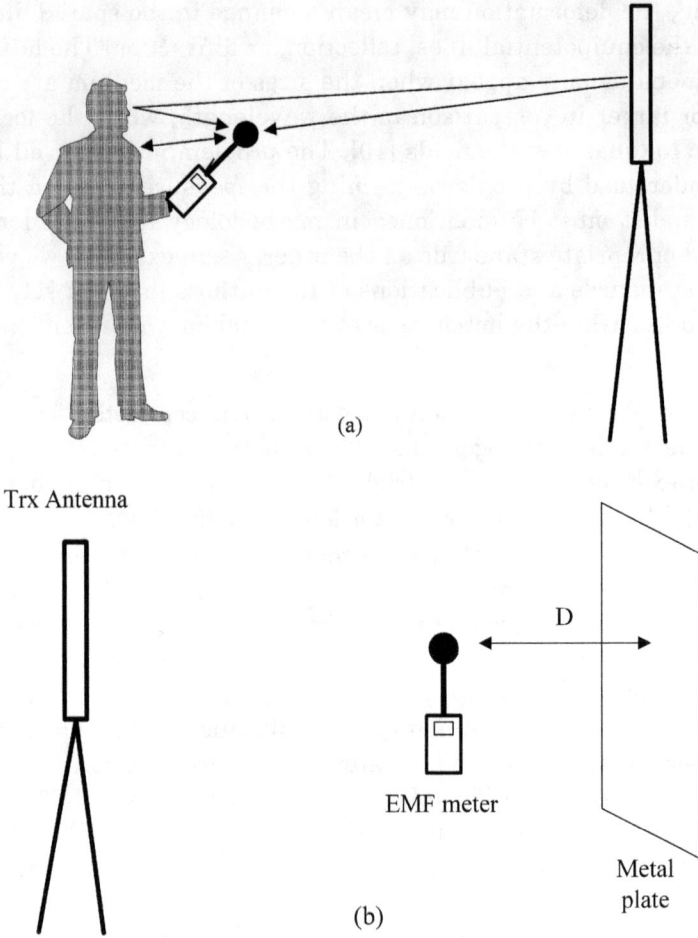

Figure 8.9   EMF reflection by (a) measuring person and (b) the experimental setup.

Figure 8.10 presents the results of measurements performed at a frequency of 900 MHz. Reflection ratio δ_{refl}, presented in the figure, is defined as follows:

$$\delta_{refl} = \frac{E_{in}}{E_{dir}}$$

(8.19)

where:

E_{in} = interfered field intensity,

E_{dir} = field of direct ray, without reflecting

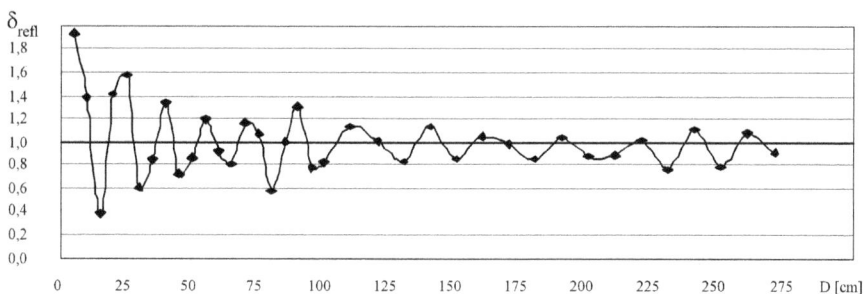

Figure 8.10   Influence of field reflection in function of distance D between probe and a conducting plate.

8.4 SUSCEPTIBILITY OF THE METER TO EXTERNAL EMF

Every electric (and especially electronic) device is susceptible to external EMF [14], and EMF meters are no exception. Because of their work in the near field and in fields of relatively high strength (unknown *a priori* as to value and spatial orientation), EMF hazard meters and similar devices must be immune to direct field penetration of the device and to high-frequency currents induced by the field in the meter's wiring. The possibilities of field penetration into a device are multiple: penetration through incomplete or inaccurate screens, dielectric slots on the junctions of varnished surfaces and those covered by other protective layers, penetration throughout metallic elements extending from the casing (leads from regulation potentiometers, switches, and indicators placed on the front panel), and indirect penetration by currents induced by the field in the probe connection cable, other interconnection cables, power cords, etc. The first designs of the hazard meters worked out at the Technical University of Wroclaw almost 40 years ago (type MPE and MPH meters) were carefully screened, and then their immunity to unwanted signals was experimentally proven [15]. An MPH-type meter shown in Fig. 8.11, designed for selective magnetic field measurement within the frequency range 0.1 to 30 MHz, well illustrates the attention paid to the screening problems at that time. These efforts have been reflected, for instance, in Polish Standards PN-77/T-06581 and PN-89/T-06580/3 in the form of a requirement to test the susceptibility of EMF hazard meters.

The meter was mounted on a standard photographic tripod that allowed rotation of its antenna while controlling it from the front panel. This made it possible to measure three spatial components of a selected frequency fringe.

Figure 8.11 MPH-type meter on a tripod.

Susceptibility reduction may be achieved via careful screening and filtering on all meter inputs and outputs. This technique is identical to those undertaken to limit electromagnetic interference in widely understood electromagnetic compatibility practices.

We should add here that the modest microelectronics, because of their relatively small size, may be seen as less sensitive to the EMF. However, there may be a pitfall here. Modern meters, placed in an elegant colored plastic cases, may be really immune at lower frequencies. But most of the time, they are unprotected against microwave fields, even though these frequencies are more and more common. The practical results are easy to foresee, and every meter user should be able to perform a simple test of the meter's susceptibility (immunity). The authors investigated examples of EMF meters with the electronics that are more sensitive to EMF than their probes. The investigations revealed that caution is necessary, especially when using inexpensive EMF indicators or similar devices. The authors are often contacted by people who complain about intentional and/or unintentional exposure to EMF. The exposure may have been detected using, for example, a smog indicator that, although designed for power line frequencies, is sensitive to wide frequency range.

The problem is illustrated in Fig. 8.12. The figure shows generalized measured frequency dependence of the susceptibility to an external EMF of the V-640-type multimeter. The measurements were performed with the use of a TEM transmission line, and the curve shows the intensity of the applied field necessary to change the multimeter's indications by 2 percent. The meter was investigated as self-powered with no external connection. During the measurement, the meter was almost insensitive to the position of the range selector switch, which may suggest that the most indicating internal portion of the meter (amplifier, detector) was affected by the field [16].

An example of an EMF meter's susceptibility to external EMF may be provided by an EMF analyzer. The device's susceptibility is shown in Fig. 8.13. The curve shows the E-field intensity required to obtain indications 20 dB above the device's noise level while an antenna of the device was disconnected and its input was loaded with a matched load. It is worth noting here that, in E-field intensities above 100 V/m, the tested device stopped working completely.

A series of measurements, performed with different types of EMF meters, showed that many types are susceptible to EMF, and penetration of the field into the devices is made via connection of the probe-indicator or an improperly screened indicator case. The latter is especially common in cheap, plastic indicators, and the effect increases when the device is held in the hand of a person performing the measurements.

These observations confirm the statement expressed at the beginning of this section and show that EMF meters designed for EMF surveying are immune to EMF. The choice of examples should not suggest to readers that this or another type of meter is better or worse. These

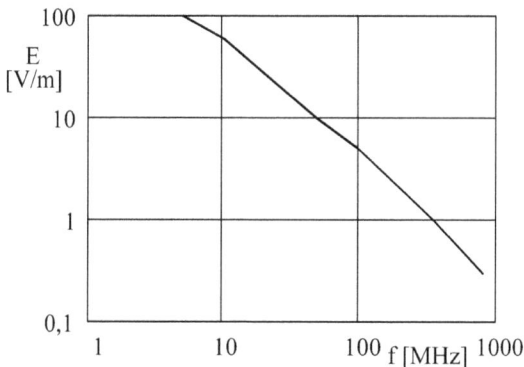

Figure 8.12 The susceptibility of a universal multimeter type V-640 to external EMF.

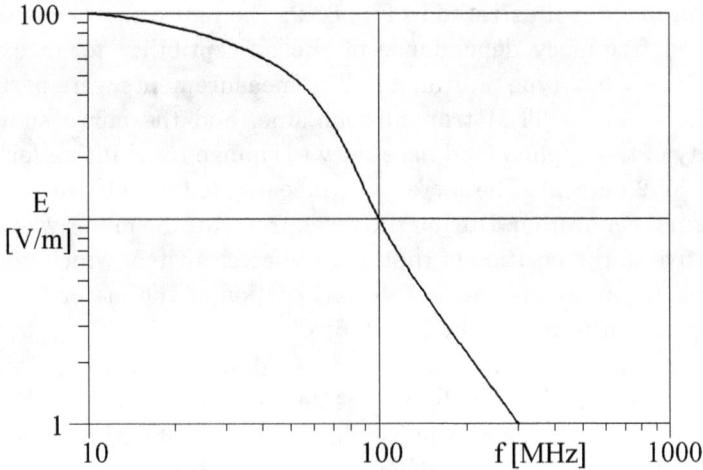

Figure 8.13   Susceptibility of an EMF analyzer.

are only examples that could be verified by a reader on his own. For this purpose, we suggest a simple test as presented in Fig. 8.14. A handheld transceiver is placed near the meter under test. It is evident that when moving the transceiver's antenna from the meter's sensor to its indicator, the indications on the meter should decrease. In several meters tested by the authors in this manner, the indications increased by as much as five times! The phenomenon was especially intense when "smog indicators" were investigated.

Several cases of meter sensitivity to ELF fields were observed. During HF field measurements in close proximity to overhead, high-voltage

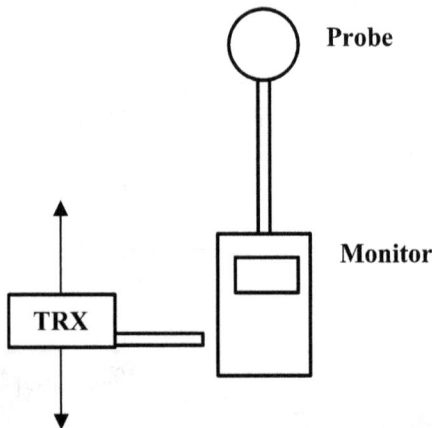

Figure 8.14   Proposed method for immunity testing at HF.

transmission lines, strange meter indications were noted. Instead of the expected indications at the level of 1 V/m, the meter shoved >100 V/m, and the indications were a function of the meter's position in relation to the line conductors, apart from the meter probe's isotropic pattern. In order to test the phenomenon a series of laboratory measurements, using several types of meters, were performed. During the experiments, a meter was placed parallel or perpendicular to E-field equipotential lines, as shown in Figs. 8.15a and 8.15b. Two metallic plates were fed from a power line frequency source (50 Hz).

Comparative results of two meters' tests are presented in Fig. 8.16. The figure allows the conclusion that the ELF E-field susceptibility of

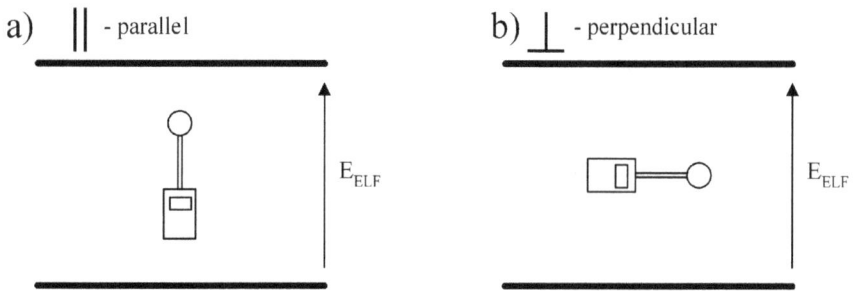

Figure 8.15   Setup for ELF susceptibility testing.

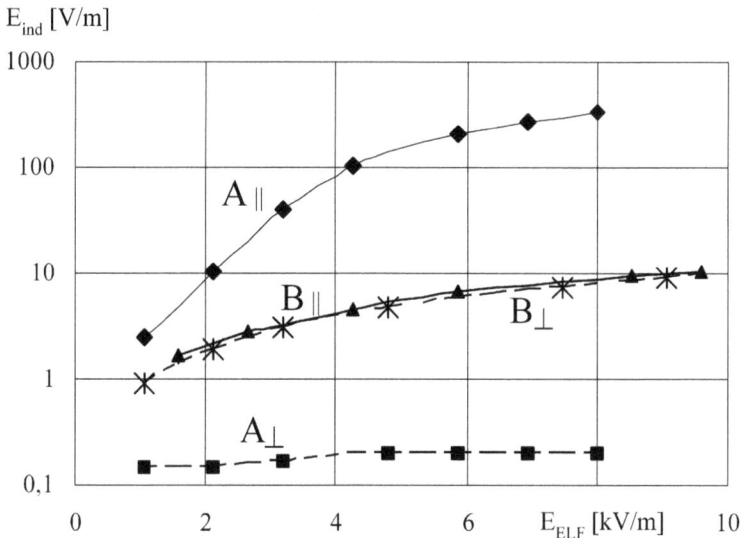

Figure 8.16   Meter's indication (E_{ind}) vs. applied ELF field (E_{ELF}) in a susceptibility test of two different RF EMF meters to the ELF field.

meter B is acceptable, while that of meter A is too large. Moreover, a difference in indications of meter A while parallel and perpendicular to the field force lines suggests that the susceptibility is due to field pickup by the connection between the meter's probe and indicating device, with no appropriate filters at the device input. The effect was observed while device A was placed near a power cord connected to household devices fed from AC power line. This suggests a simple procedure by which users can test meters.

8.5 RESONANT PHENOMENA

Deformations of the measured field caused by the meter, its operator and, at the least, any conducting medium were presented above. In addition to these discussed phenomena, especially in the near field, resonant-size objects may affect the field distribution around them more intensely than nonresonant ones. The resultant field around these objects may significantly exceed the primary field. We will refer to this phenomenon as *field amplification* by a resonant object [17]. These objects play a role as secondary passive radiators. The use of passive retranslators is well known in radio communication, and a good example of their use is the Yagi-Uda antenna in which only one element (radiator) is excited from a source while the others (reflectors and directors) are excited by the field of the radiator. Because of their resonant sizes, they are able to focus the radiated energy. Figure 8.17 plots the results of an estimated amplification magnitude in the neighborhood of a passive half-wave dipole, in the form of multipliers $n(\rho)$, which should

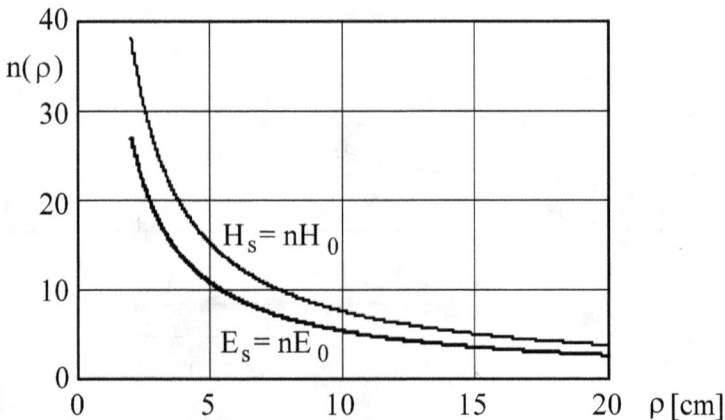

Figure 8.17   EMF amplification of the half-wave passive dipole at 150 MHz.

be applied to the intensity of the primary electric field E_0 and the magnetic field H_0 to obtain the intensity of the secondary electric field E_s and the magnetic one H_s in their maxima, i.e., the electric field near the dipole ends and the magnetic field near its center. The chosen dipole resonant frequency was 150 MHz. However, because of the linearity of the phenomenon, these magnitudes of amplification may be transferred to any other half-wave resonant dipole by multiplying the amplification factors read from the curves by $150/f_x$ (where f_x = the resonant frequency of the dipole in megahertz).

As may be seen from Fig. 8.17, at a distance of 5 cm from the ends of the resonant passive dipole, the intensity of the electric field increases ten times, whereas that of the magnetic field (at the same distance from the dipole center) increases fifteen times! Although the effect decreases with frequency, even up to frequencies of, say, 1 GHz, it is substantial enough that it should be taken seriously.

The measured resonant frequencies of different objects used in everyday life are shown in Table 8.1.

Table 8.1   Measured Resonant Frequencies of Commonly Used Objects

Object	f_{res} [MHz]	Object	f_{res} [MHz]
Table	30–60	Tools	300–500
Chair	70–100	Cutlery	300–600
Stool	100–150	Glasses	350–1000

To illustrate the scale of the field deformations caused by resonant phenomena, as well as to show the magnitudes of the field amplification by the resonant secondary radiators, Fig. 8.18 shows a standing wave within a metal door frame at an FM station frequency of around

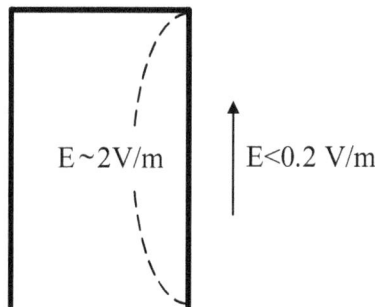

E ~2V/m   E<0.2 V/m

Figure 8.18   The standing wave in a metal door frame.

70 MHz (in this case, the standing wave distribution was used to identify the source of radiation). Figure 8.19 shows the field distribution around a metal kitchen stool at its resonant frequency, near 135 MHz.

In every investigated case, the field amplification significantly exceeded 20 dB at a distance 5 cm from the measured object. The effect was extremely strong in the case of eyeglasses with metal frames while portable transceivers were in use [18]. The frames not only confirm the role of the resonant phenomena and amplify the field around them, they also concentrate the field around the device user's head. What is not taken into account in the data presented in Table 8.1 is the objects' influence on resonant frequencies and their amplification in proximity with other material media. In the case of the spectacles, for instance, their resonant frequency is much lower while being worn by a person than when just sitting in free space; the field distribution around them, of course, is different from that around a linear dipole, and their arms form something like capacitor plates, causing the above-mentioned field concentration phenomenon inside one's head.

The resonant phenomenon requires not only special care during close field measurements but also a modification of the protection standards with regard to the distance at which a measurement should be performed. The example of the eyeglasses best supports the point.

As may be seen from the above considerations, the resonant frequency of a device is a function of its size and shape, and the frequency depends on the presence of other nearby material bodies (human head, hand, or some kind of support). It may be a good idea for the reader to determine the resonant frequency for a particular device and the devices surrounding it. For the purpose, we may apply a set as shown in Fig. 8.20. An object under test (OUT) is illuminated from a source, in this case a log-periodic antenna (or other type of wideband antenna) fed

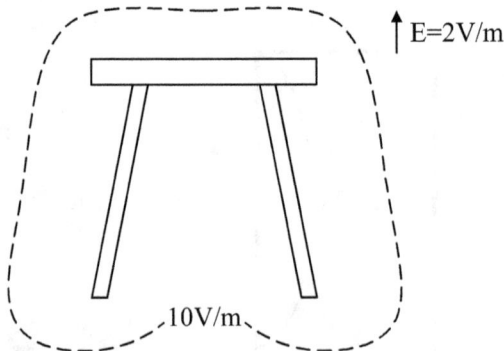

Figure 8.19   EMF around a kitchen stool.

Figure 8.20   Setup for resonant frequency measurement.

from a generator (G). The device is coupled with transverse component of the field. Near the device is placed a short dipole sensor (DS) such that it should be sensitive to the radial component of secondary EMF generated by the device. For this purpose, it is most convenient to use a probe with sinusoidal pattern (see Section 7.4.1). The component shows its maximum at the resonant frequency of the device.

8.6   HUMAN FACTOR

Human factor is defined by the authors as the influence of skills, experience, and excellence of a person performing measurements on the measurement results. The factor applies during any metrological procedure, but here it may play a basic role, especially in survey measurements, which are typically performed in specific and varying conditions of time and space that often are not repeatable. The measurements may be performed with the use of different measuring devices whose parameters may be not comparable, which leads to remarkable differences in measurement results. Sometimes, especially in small laboratories, the measurements are the "by-product" of specialists from other environmental pollution areas, and this may lead to errors and mistakes due to a lack of understanding of electromagnetic phenomena. To demonstrate the role of the human factor, a series of comparative measurements were performed. The measurements were performed in similar conditions and by personnel experienced in the area. The first comparative measurements were performed in laboratory conditions while a group of surveyors, in consideration of repeatability conditions, measured E-fields from the same source, in the same points, using the same measuring devices. This is similar to NIST requirements [19]. The results of these measurements confirmed that the factor may play significant role [20].

In similar conditions, except at an outdoor location, measurements were performed in an interlaboratory comparison program. Well experienced representatives of several Polish survey labs took part in the comparison. During the measurements, E-field stability was monitored to assure a constant level to within ±3 percent. The comparison revealed that the measurement results reflected only "pure" human factors, i.e., the ability to discern the readouts of a meter. Figure 8.21 presents the results of field intensity readouts performed by 18 persons.

To better illustrate the problem of the human factor, which in our presented measurement can be defined as "repeatability of measurements by the measuring person," Fig. 8.22 presents comparison results of three series of measurements in which 18, 19, and 20 persons took part. All series were independent of each other and performed in three separate sessions. The human factor (HF) indicated in the figure is defined as the percentage (%) deviation of a person's readouts in relation to mean value. The figure shows the deviations in increasing order from the most negative to the most positive, and participants were

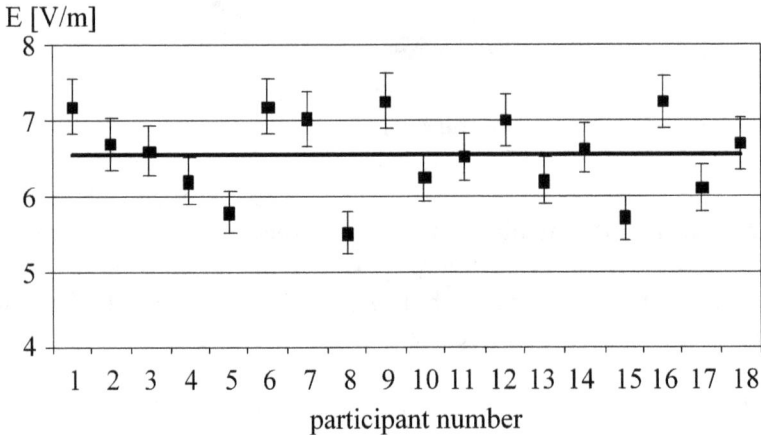

Figure 8.21 Measurement results from an experiment.

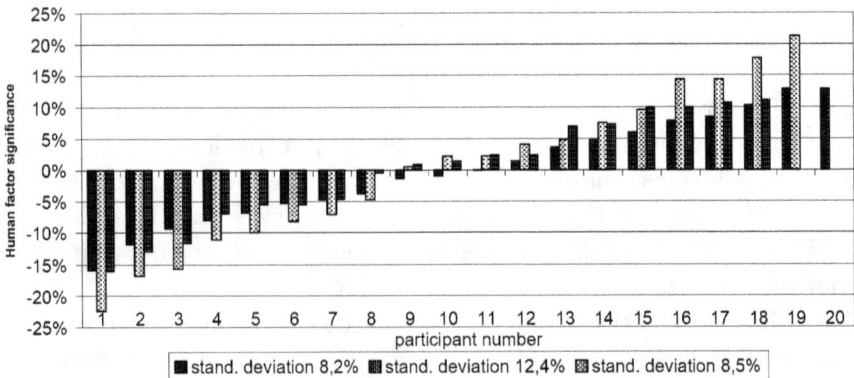

Figure 8.22 A statistics of three series of measurements.

numbered arbitrary in each series of measurements. In the figure are indicated standard deviations of each series, i.e., 8.2 percent, 12.4 percent, and 8.5 percent. The figure leads to the conclusion that the uncertainty of EMF measurements due only to the human effect may be estimated on the level of ±10 percent, while the maximum dispersion may exceed ±20 percent. According to an EC Recommendation [21], it is possible here to use the A-type uncertainty evaluation method, which leads to mentioned ±10 percent standard deviation using rectangular or normal distribution law.

8.7 UNCERTAINTY BUDGET

Two introductory comments about uncertainty are:

- This is a typical approach; sources of uncertainty shall be identified and taken into account in accuracy estimations of any metrological procedure. The estimations should be repeated for any series of measurements or when measurement conditions have changed.

- The same requirement is valid in relation to EMF measurements. However, time and space variability and other factors limiting the measurement accuracy may lead to a pessimistic statement that the uncertainty of the EMF measurements, especially in the discussed area, may lead to a change from quantitative investigations to qualitative ones.

Disregarding the pessimistic statement, we should at least be able to say something about the accuracy of performed measurements. The procedures for uncertainty estimations, generally understood in electromagnetic compatibility (EMC) practices, are well known from the literature, and they are specified, for example, in Ref. [21]. They are similar in our case, so we won't discuss them in detail. We would like only to note that, in this area, these measurements are performed quite often for legal reasons. Thus, measured EMF levels may be in agreement or disagreement with national (or international) reference levels. In the both cases, the measurement uncertainty should be determined at the same time as the compliance level assessment, as shown in Fig. 8.23 [21].

Different standards, regulations, and metrological literature require different approaches to the total uncertainty estimation. The basic rule here is to identify separate factors that limit accuracy and so it is possible to evaluate their levels and the roles they play in measurement accuracy degradation. One of the most universal guidelines for uncer-

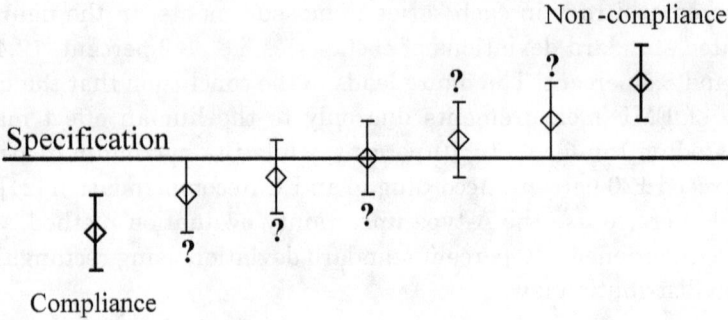

Figure 8.23 Typical cases of compliance assessment.

tainty evaluation is the *ISO Guide to the expression of uncertainty in measurement* [22]. In EMF measurement, both A and B types of evaluation presented therein shall be used to estimate individual components of total uncertainty. The combined (total) standard uncertainty u_c is calculated by the RSS (root-sum-square) method:

$$u_c = \sqrt{\sum_{i=1}^{m} c_i^2 \cdot u_i^2}$$

(8.20)

where:

u_i = standard uncertainty of i- component,

c_i = sensitivity coefficient (typically, $c_i = 1$).

Because the standard uncertainty assures a confidence interval of 68 percent, and most assessment methods require a higher confidence level, the expanded uncertainty must be used. Expanded uncertainty is defined as $u_e = k \cdot u_c$, where k is a coverage factor. The typically used coverage factor $k = 2$ assures the possibility of obtaining a confidence level of 98 percent.

Although it is impossible to prepare a universal recipe for EMF measurements in uncertainty estimation, Table 8.2 shows uncertainty levels estimated by the authors that focus on the most important factors. These levels illustrate and summarize the considerations presented, and they may be helpful to readers in their practice.

The table mainly takes into account factors in selected and stable space-time measurement conditions. Although an influence of a source's instability or the presence of interfering sources is indicated, especially while surveying related to human safety or environment protection purposes is performed, this is not enough. In the latter case, it may be

Table 8.2 Uncertainty Budget

Source of uncertainty	Chapter	Probability distribution [a]	Divisor	Typical value/comments
1	2	3	4	5
Measurement method	4, 5, 6	N	1 or k	0.5–1.0 dB, the uncertainty increases with distance probe-source decrease due to probe-source couplings
Measuring equipment				
Calibration		N	1 or k	0.5–3 dB, expanded uncertainty with k = 2, the uncertainty increases with frequency increase; note: usually calibration is performed in cw EMF
Linearity		U	$\sqrt{2}$	0.2–2 dB, well designed meter should have linearity below 0.5 dB (with internal corrections or calibration curves)
Isotropy	7	R	$\sqrt{3}$	0.5–3 dB, a role of the factor may be minimized by the probe rotation to max indication; note: isotropy may be a function of frequency and field intensity
Frequency response	4, 5	R	$\sqrt{3}$	0.5–1 dB, in calibration points in frequency range up to 3 GHz, 1–3 dB or worse above 3 GHz
AM modulated fields	8	R	$\sqrt{3}$	0–2 dB for full dynamic range, up to 0.5 dB for RMS detection (typically 15–25% for full dynamic range)
Pulsed field (pulse duration< 1%)	8	R	$\sqrt{3}$	Stationary field, 0.5 dB for thermocouple detector, 1–4 dB for diode detector; >10 dB for nonstationary fields, may be decreased by use a dedicated calibration factor or appropriate calibration
Temperature and humidity	8	R	$\sqrt{3}$	0.01–0.07 dB/K for humidity <75%, depending of measurement time in the conditions due to the thermal inertia of the meter
Meter reading error and indication's fluctuations	8	T	$\sqrt{6}$	< 0.5 dB; note: warming when peak or max hold are used–this could be a source of gross measurement's error caused by electrostatic charges, and field source power fluctuations
Influence of fields out of measurement range	4, 5, 6	N	1 or k	Depends on measurement conditions: 0 dB in monochromatic field, 1–3 dB in complex fields; may be limited by checking EM environment and switching off interfering sources
Susceptibility to unwanted field component (E in H measurement and H in E measurements)	5	R	$\sqrt{3}$	<1 dB, usually a sensitivity to unwanted component is well limited, however, a caution is necessary while measurements near a source of dominating E or H field; e.g., H-field near inductive heaters
Human factor	8	N	1 or k	<1–2 dB for experienced people
Drifts in field generated by source		R	$\sqrt{3}$	Negligible for stable sources and laboratory conditions, 3 dB for time-varying fields and *in situ* measurements; e.g., near welding machines

[a]Probability distribution: N = normal (Gaussian), U = U shape, R = rectangular, T = triangle.

necessary to investigate the field variations in time (dose measurement), appropriate selection of measuring point, and other factors that are loaded with errors specific to them.

If we take into consideration all factors degrading measurement accuracy, one can say that near-field EMF measurements and EMF surveying may be loaded with errors within the region of ±2–6 dB or worse. In the majority of cases, the accuracy degradation is a result of objectively existing factors and reasons, and this accuracy level must be accepted as satisfying our goals. Let us note that uncertainty at the level of ±2 dB, in the light of considerations presented here, may be achievable at lower frequencies, in stable cw fields, when measurements are performed by experienced personnel in steady conditions, using good quality and well calibrated meters. Any change in conditions leads to accuracy degradation.

References

[1] C. Dragone. Performance and stability of Schottky barrier mixers. *The Bell System Techn. Journal,* vol. 51, No. 10/1972, pp. 2169–2196.

[2] N. Inoue, Y. Yasuoka. Responsivity of antenna-coupled Schottky diodes. *Infrared Physics,* vol. 25, No. 4/1985, pp. 599–606.

[3] G. Gerbi, D. Golzio. A new EM field sensor for radiation hazard and EMI measurements. *Proc. IEEE 1983 Intl. EMC Symp.* Arlington, pp. 142–146.

[4] K. A. Chamberlin, J. D. Morrow, R. J. Luebbers. Frequency-domain and frequency-difference, time-domain solutions to a nonlinearly-terminated dipole: theory and validation. *IEEE Trans.* vol. EMC-34, No. 4/1992, pp. 416–422.

[5] *Geraetedokumentation Nahfeldstarkemessgeraet* NFM-1 (in German), Haidenau, 1986.

[6] R. G. Harrison, X. Le Polozec. Nonsquare law behaviour of diode detectors analyzed by the Ritz-Galerkin method. *IEEE Trans.* vol. MTT-42, No. 5/1994, pp. 840–846.

[7] T. M. Babij, H. Trzaska. Quarterly reports for the NBS grant No. NBS(G)-176 (unpublished). Wroclaw, 1975.

[8] B. R. Strickland, N. F. Audeh. Numerical analysis technique for diode-loaded dipole antenna. *IEEE Trans.* vol. EMC-35, No. 4/1993, pp. 480–484.

[9] M. Kanda. Analytical and numerical techniques for analyzing an electrically short dipole with a nonlinear load. *IEEE Trans.* vol. AP-28, No. 1/1980, pp. 71–78.

[10] E. Grudzinski, H. Trzaska. EMF indicators. 1984 Annual Meeting of the BEMS, Copenhagen.

[11] H. D. Bruens, H. Singer, T. Mader. Numerical investigations of field distortions due to sensors. *IEEE Trans.* vol. EMC-35, No. 2/ 1993, pp. 110–115.

[12] J. D. Norgard, R. M. Sega, M. Harrison, A. Pesta, M. Seifert. Scattering effects of electric & magnetic field probes. *IEEE Trans.* vol. NS-36, No. 6/1989, pp. 2050–2057.

[13] T. Bossert, H. Dinter. Beurteilung der Gefaehrlichkeit starker elektromagnetischer Felder-kalibrierung und messgenauigkeit von Nahfeldsonden im Berich 30 kHz–30 MHz (in German). *JTG Fachbericht* No. 106/1988, pp. 57–62.

[14] H. Cichon, H. Trzaska. Selected susceptibility problems of the general use electronic equipment. *Proc. 1979 Intl. EMC Symp.* Rotterdam, pp. 123–126.

[15] T. M. Babij, H. Trzaska. Selective EMF meters (in Polish). IME Publications of the Technical University of Wroclaw No. 1/1970, pp. 55–66.

[16] H. Cichon, H. Trzaska. Susceptibility problems of home entertainment electronic devices. *Proc. 1981 Intl. EMC Symp.* Zurich, pp. 295–299.

[17] H. Trzaska. Resonant phenomena and their role in dosimetry and protection standards. XVII-th Annual Meeting of the BEMS, Boston, 1995, p. 188.

[18] H. Trzaska. EMF near passive secondary radiators. USNC/URSI Radio Science Meeting, Newport, CA, p. 59.

[19] Barry N. Taylor and Chris E. Kuyatt. Guidelines for Evaluating and Expressing the Uncertainty of NIST Measurement Results, NIST Technical Note 1297, 1994.

[20] P. Biekowski, H. Trzaska: Interlaboratory comparisons in EMF survey measurement—methods and results. International Conference and COST 281 Workshop on Emerging EMF Technologies, Potential Sensitive Groups and Health, Graz, Austria, 2006.

[21] Guidelines on Assessment and Reporting of Compliance with Specification (based on measurements and tests in a laboratory) ILAC-G8: 1996.

[22] Guide to the expression of uncertainty in measurement (GUM), ISO/IEC Guide 98: 1995.

Chapter 9

Photonic EMF Measurements

Until now, the dominating technique for EMF measurement was to use an antenna (mainly a dipole or a loop) loaded with a detector (a diode or, more rarely, a thermocouple) and transfer the DC voltage from the probe to an indicator (in the case of the most popular designs of two-piece meters) through a high-resistance (transparent) transmission line (Fig. 9.1a). The most important inconvenience of this technique is the vanishing of phase information that is (sometimes) indispensable; similarly, the spectral information is lost as well. Although the latter is usually unnecessary, especially where wideband measurements are of concern, we will see in our further considerations that it is possible and advantageous to perform wideband measurements using spectral information as well.

Among all considered media for information transmission, the photonic one seems to be, in our field of involvement, the most promising. Only recently have proposals been worked out for photonic field meters, and the first designs are being investigated and even applied under laboratory conditions. This technique is often employed for data transmission and for remote reading of measured values. An example of such a meter design is shown in Fig. 9.1b.

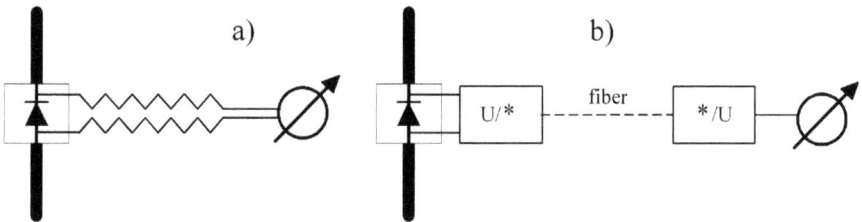

Figure 9.1 EMF meters: (a) traditional design with a resistive transmission line, and (b) design with optical transmission.

The advantages of this concept may be well illustrated by the EMF meter designed at the Technical University of Wroclaw in the 1970s and used mainly for EMF measurements in proximity to transmitting antennas. The measurements were and are performed for experimental verification of theoretical hypotheses on the spatial EMF distribution around transmitting centers when their modifications or reconstructions were planned, in order to evaluate exposure to the nearest inhabited areas. Similar measures are done when new buildings are planned in the neighborhood of the centers. According to Polish environmental protection standards, in these situations, the theoretical estimation of the exposure, in relation to the limits provided by the standards, is an initial requirement for investment approval. Then, after the investment is completed, the estimations should be confirmed by measurements.

An EMF meter suitably modified for this purpose, equipped with a probe adequate for the selected frequency range and to the measured component of the field, was hung under a weather balloon. The output voltage from the meter was converted into an optical signal and transmitted to a demodulating device on the ground via an optical fiber and then to a recorder (Fig. 9.2) [1,2].

Contrary to any other wire or wireless transmitting media, the presented solution is absolutely insusceptible to external interference, and what is especially important is that it is insulated against unwanted influences of the measured field. (This, by the way, is presently a source of many EMC-related problems, and in particular the remarkable sensitivity of any type of electric or electronic device directly exposed to an

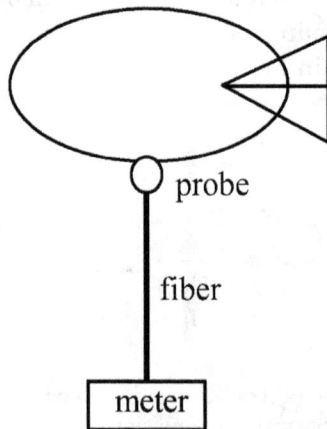

Figure 9.2 Spatial EMF distribution measurement with the use of a captive balloon.

external EMF or to voltages induced by the field in the device's wiring and interconnection cables. This is mentioned in Section 8.4.) The optical fiber used here for data transmission (used simultaneously as the captive line of the balloon) does not cause any disturbances in the measured field. The flying part of the measuring system is insulated galvanically against the meter situated on the ground, which ensures that the use of the system is absolutely safe in terms of the possibility of the measuring team being shocked by HF currents, power-line frequency currents, or static discharges. It allows us to perform the measurements in the close proximity to big power antennas, their guy wires, towers, overhead power lines, and other charged devices.

Although this construction has many advantages (some of which are listed above), it is a relatively "primitive" one, since the photonic approach is here used "only" to enable remote readings of the measured data. As already mentioned, similar designs (with considerably reduced transmission distances and other requirements) are presently offered by many manufacturers in the form of so-called *repeaters* that could be connected to EMF meters and indicators. The subject of our concern in the ongoing part of the work will be the possibility of using of photonic elements as sensors (transducers) of the measured field. The advantages of also using the photonic technique here for data transmission will be, in some sense, a "by-product" of the method.

Several introductory comments are in order:

1. There is abounding literature devoted to the field of optoelectronic elements' applications as sensors for measurement of a variety of physical quantities. A number of publications are concerned with optoelectronic transducers use in EMF measurements [3, 4]. Because of this, we won't present any particular solutions in detail, and we will concern ourselves with only a few of the ones that are especially useful for our applications in terms of illustrating the tools and methods that, in all likelihood, will dominate in the near future.

2. The principle of a photonic transducer may be conveyed to the modulation of an optical beam, and the subject of the modulation may be the phase of the signal (which is a function of the velocity of the light propagation in an electro-optic medium), its frequency, amplitude, or polarization. The type of modulation, as well as that of the electro-optic crystal, are selected so as to obtain the maximal sensitivity of the device. In the role of coherent light sources, lasers are usually used—most readily, monomode semiconducting lasers.

3. While current literature discusses the sensitivities of photonic meters, which exceed that obtainable using traditional techniques,

sensitivity is still the Achilles' heel of the photonic technique, and its increase is the subject of technological efforts (i.e., the development of electro-optical materials with greater sensitivity than presently used ones) as well as technical ones (i.e., the development of electro-optic systems and electronic circuits that will allow an increase in sensitivity while assuring the required stability and reasonable prices).

4. In the application discussed herein, in general, two approaches are used:

   • A direct interaction of the measured field upon the electro-optic crystal.

   • A voltage (or rather an emf) induced by the field in an auxiliary antenna (playing the role of the measured field concentrator) impressed to the modulator.

   The latter is usually used to increase the sensitivity of the meter by way of the application of relatively large-size antennas. Although, in the former approach, the large-sized sensors (for a sensitivity increase) may be applied as well, it results in a reduction in the stability of the system. In the latter, analyses of the measuring band (including the necessity of using RC frequency response shaping filters) are identical to that presented in Section 4.2 [5]. In the former, there is no lower corner frequency (it also allows measurement of the static fields), while the upper bandpass limit depends only on the properties of the electro-optic material applied in the modulator and its size.

5. Apart from the technological differences, the factors limiting accuracy of the measurement are, in the case of photonic probes, similar or identical to those presented in previous chapters. They may be related to the frequency response of the probe, which was mentioned above, its size (not taking into consideration insignificant interaction of a source with the properties of the probe in the version with no antenna), deformation of the spherical directional pattern, susceptibility to temperature variations, and factors specific for the photonic technique; for instance, instability in the power generated by a laser, optical power losses, modulator heating, and others.

9.1 THE PHOTONIC EMF PROBE

Contrary to the design shown in Fig. 9.1b, in a photonic sensor, the measured EMF is used for light generation (Fig. 9.3a) in a probe with optical modulation of a coherent beam of light from a laser (Fig. 9.3b). As mentioned above, the modulator may be affected directly by the measured field, or (for purposes of sensitivity increase) an antenna (of electric or magnetic type) additionally may be applied.

In photonic EMF sensors, it is possible to modulate the beam in amplitude, phase, or polarization. Although the final result of the procedure is always amplitude modulation, direct amplitude modulation is rarely applied because of its low sensitivity and the resulting poor signal-to-noise ratio. Block diagrams of photonic EMF meters using polarization modulation and phase modulation are shown in Fig. 9.4.

Both solutions shown in Fig. 9.4 have similar parameters. Thus, for illustration, we will discuss one of them in a more detailed way: the probe with the Mach-Zehnder interferometer. The light from the laser is split in the divider into two parts. One of them becomes the reference beam, whose phase (propagation velocity) should be independent of the measured field, while the other, due to variation in the modulator's permittivity as a function of its exposure to a field, and as a result the phase velocity alternations (modulation) of the optic ray, is phase modulated. Then, the beams are interfered in the adder, which results in amplitude modulation of the primary light beam at the output of the interferometer.

The output beam is led to a detector, and then the signal is analyzed by a measuring device. The modulation depth of the signal, in a given

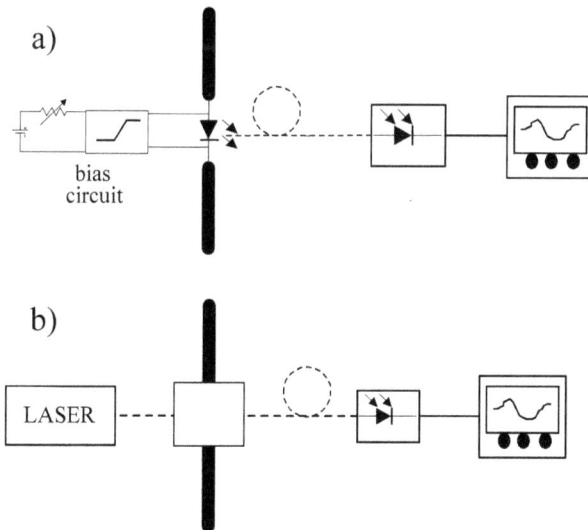

Figure 9.3 (a) Block diagram of a photonic EMF meter with an optical conversion and (b) laser beam modulation.

a)

b)

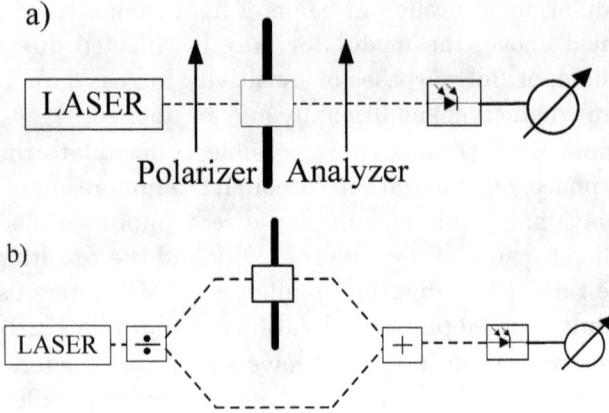

Figure 9.4 Two types of photonic EMF meters: (a) polarimetric and (b) interferometric.

range, is linearly dependent on the amplitude of the measured field, while the frequency of the signal (including its envelope) is exactly equal to the modulating frequency. This statement concerns every frequency fringe of the signal that is led to the modulator from the antenna (measured field). Because of the modulator linearity, unwanted products of intermodulation, a distinctive feature of active antennas, should not appear at the detector's output. The latter well illustrates the role of the presented device not only for measurement purposes but as an active receiving antenna as well. In the device, the intensity of the optical signal at the detector (I) is given by:

$$I = \frac{I_0}{2} \left[1 + \sin\left(\pi \frac{V}{V_{\lambda/2}} + \varphi_0 \right) \right]$$

(9.1)

where:

I_0 = light beam intensity at the output of the modulator,

V = modulating voltage,

$V_{\lambda/2}$ = the voltage causing phase difference in the modulator in π,

φ_0 = introductory phase.

The introductory phase results from an inaccuracy in the optical tracts manufacturing and alignment. Because of its importance for the adjustment of the modulator's operating point, the modulator setup requires precision work. It is required to be able to control the operating point without requiring mechanical interference with the device. The role of the introductory phase is illustrated in Fig. 9.5, using Eq. (9.1) to calculate the output signal of the interferometer for $\varphi_0 = 0$ and for $\varphi_0 = \pi/2$ [6].

It should be observed here that the solution shown in Fig. 9.5b is more beneficial both because of its relatively large range in the linear part of its dynamic characteristics and its higher sensitivity. In both cases, the characteristics are periodic and, as a result, if the modulator input signal exceeds a certain value, the output voltage of the detector not only does not increase but may even decrease.

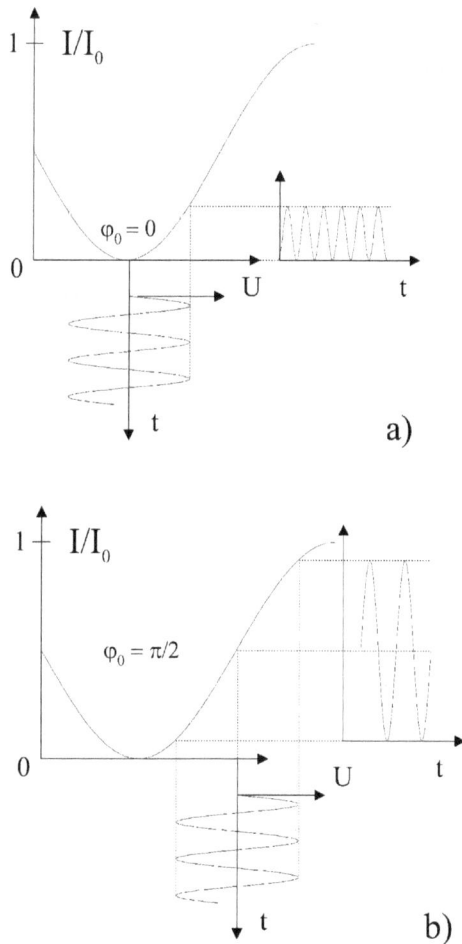

Figure 9.5 Dynamic characteristics of an interferometer: (a) $\varphi_0 = 0$ and (b) $\varphi_0 = \pi/2$.

9.2 FREQUENCY RESPONSE OF THE PROBE

Although above it was ascertained that there is extensive similarity between frequency response analyses of a traditional probe and those of the photonic one, some features of the response, especially those distinctive to the photonic probe that may be useful for emphasizing its advantages, require a more accurate approach and a focus of our attention on differences rather than similarities.

To shorten our considerations here, we will refer the reader to those in Chapter 4 and assume, without additional explanations, that, much as in the case of traditional probes, with photonic probes the measuring band limitation, in the range of the highest frequencies, is important here. Figure 9.6 shows the structure of a probe with an optical modulator, without an RC low-pass filter and with such a filter. An equivalent network of the probe with the low-pass RC filter is shown in Fig. 9.7.

Using certain modifications of the equivalent network shown in Fig. 9.7 in relation to those applied in Fig. 4.2, the transmittance of the

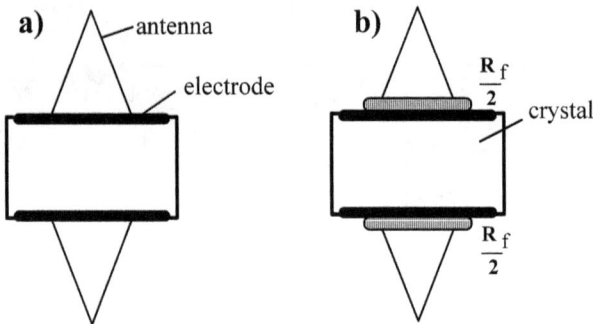

Figure 9.6 Structure of a photonic EMF probe (a) without RC filter and (b) with the filter.

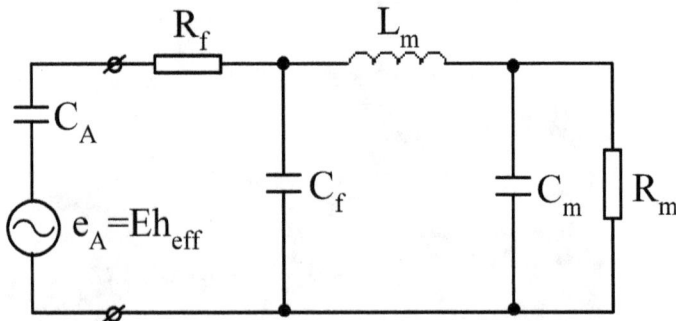

Figure 9.7 Equivalent network of the probe with the RC filter.

probe shown in the former is given by a formula identical to Eq. (4.12) while we assumed $L_f = 0$. Let's direct our attention to the specific differences in the transmittance's frequency run for both the traditional probe and the photonic one.

A. The Lower Corner Frequency

The lower corner frequency f_l of the photonic EMF probe, at which its transmittance diminishes by 3 dB in relation to the value within the measuring band, using the indications as in Fig. 9.7, may be defined in the form:

$$f_l = \frac{1}{2\pi R_m \left(C_A + C_f + C_m \right)}$$

(9.2)

Equation (9.2) is identical to Eq. (4.15); however, it requires some comments.

a. If the measuring antenna is loaded with a detection diode, as a result of the diode's nonlinear properties, both its internal capacitance and the detector equivalent resistance are functions of the current flowing through the diode. Thus, both the lower corner frequency and (to a lesser degree) the transmittance within the measuring band depend on the measured field strength, which has already been examined in Chapter 8. In the case of the photonic probe with phase modulation, the modulator equivalent capacitance C_m also varies as a function of the applied voltage (EMF), because the modulation is realized through ε_r variations in the voltage function, but the dependence is much less that that of diode detectors. On the other hand, the modulator equivalent resistance R_m is voltage independent. Thus, the stability of the lower corner frequency and the absolute magnitude of the transmittance within the measuring band are, in photonic sensors, much higher than those of traditional ones. The comparison is much more advantageous for the photonic probe in terms of its thermal stability; it is sensitive to the temperature variations as well, but in its optical part.

b. The acquisition of the possibly low f_l in the traditional probe requires careful selection of the type and, even, specimen of the detection diode to obtain a high value of R_d. In the photonic probes, high values of R_m are the rule. This makes it possible to achieve a lower corner frequency well below that of traditional probes.

c. It has already been mentioned that there exists the possibility of directly using an optical modulator in the role of the field probe. In this case, the antenna capacitance C_A does not exist. If we assume in Eq. (9.2) that $C_A \to \infty$, we will have $f_l \to 0$! The sensitivity of such a probe won't be very high indeed. However, field measurement starting from static fields is so attractive that this possibility is the subject of considerable investigation. We should observe here that the statement may be applied to both the electric and magnetic fields. It should be noted that, in a magnetic field sensor, static field measurement is possible only in the version using a modulator with a magneto-optic crystal. In the version with a loop antenna and an electro-optic modulator, the lower corner frequency is similar to that of probes with loop antennas, which were discussed in Chapter 5; a similar result, in much narrower frequency range, may be reached using a Hall cell. In the case of an electric field probe, the static field measurement is not possible by any other currently known measuring method.

B. The Upper Corner Frequency

The upper corner frequency f_u of the probe presented in Fig. 9.7 is given by:

$$ f_u = \frac{C_A + C_f + C_m}{2\pi R_f C_A \left(C_f + C_m \right)} $$

(9.3)

The structure of the photonic probe shown in Fig. 9.6 has a much lower modulator inductance L_m compared with to the parasitic inductances of traditional probes, even those realized using a thin-film hybrid technology [7]. As a result of this, the filter capacitance C_f and the modulator capacitance C_m should be considered as a single element—let's call it the resultant capacitance of the modulator C'_m. Consequently, the self-resonance of the modulator will appear at a frequency much higher than similar resonances in traditional probes, and the time constant of the $R_f C_f$ filter can be adequately smaller in relation to the considerations of Chapter 4. In the no-antenna probes, the filter limiting the upper corner frequency is not necessary, and the measuring band of the probe is limited in the highest frequency range by the maximal sizes of the probe. Furthermore, it is defined, for example, by the permissible value of the errors δ_{1E} and δ_{2E} (which were estimated in Chapter 4), or by analogous limitations of the electrical and the geometrical sizes of the magnetic field probe, (which were considered in Chapter 5). In the latter case, in relation to the no-antenna probe, the antenna effect does not exist, without regard to a certain sensitivity of the photonic electric field probe to the magnetic field, and *vice versa*. It is worth remembering that, in the case where the upper frequencies are unlimited artificially, the transmittance in this range may be arbitrary. As a result, measuring errors may be arbitrary as well.

Apart from the above-discussed factors limiting wideband characteristics of the photonic probes, our attention must be focused here on a very important factor: the simultaneous appearance of piezoelectric phenomena in electro- and magneto-optic crystals. Then, when the modulating frequencies are near those of the piezoelectric resonances, a remarkable increase in modulation efficiency, resulting from coincidence of the two phenomena, will appear. These resonances do not exclude the possibility of using such a crystal in an optical modulator at frequencies above the first piezoelectric resonance; however, the frequency response of the modulator above the frequency will have a comb-like character. This means that, at harmonics of the basic resonant frequency and at harmonic frequencies of any other crystal oscillations' modes, there will appear strong and narrow maxima. Sometimes in metrological applications it could be enough to precisely know all these frequencies and, as a result, the shape of the frequency response of the modulator (probe), then they may be taken into account during computer analysis of the measurement results. However, the resonances strongly depend on temperature, which makes the analysis inaccurate. It results in the trend to use EMF probes with a flat frequency response within the measuring band, in which case the upper part of the response should be eliminated artificially (for instance, by the use of the above-mentioned RC low-pass filters). In one designed and investigated modulator model using an $LiNbO_3$ crystal, the resonant effects are very strong; its sensitivity at the resonant frequencies exceeds several times that of the sensitivity outside them. The phenomenon is well illustrated by the measured frequency response of the hybrid modulator (Fig. 9.8) [7].

In a case where the frequency band in which the resonances appear should be artificially eliminated, the first resonant frequency and, as a result, the upper corner frequency of the probe may be estimated as:

$$f_u \approx \frac{v}{2\,d}$$

(9.4)

where:

v = velocity of the acoustic wave propagation in the crystal (usually 4–5 km/s),

d = transversal size of the modulator.

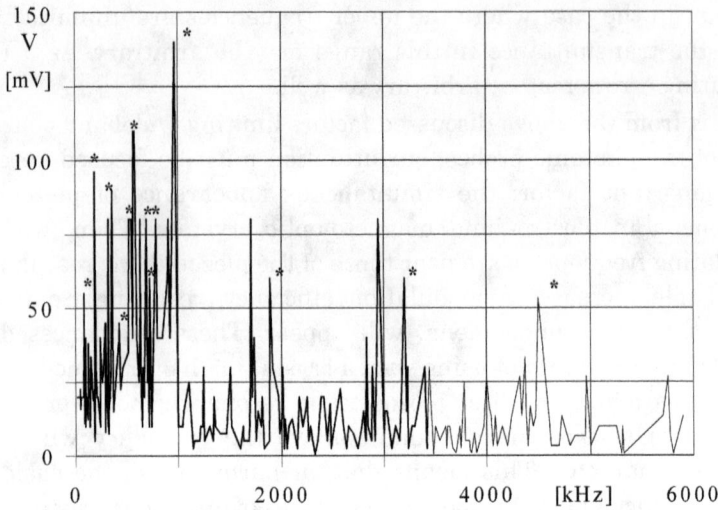

Figure 9.8   Measured frequency response of LiNbO$_3$ modulator.

9.3 SENSITIVITY OF THE PHOTONIC PROBE

The detector output voltage V_d in the Mach-Zehnder interferometer with an optimally selected working point, i.e., for $\varphi_0 = \pi/2$, for the medium frequency range is given by:

$$V_d = \frac{C \cdot I}{2} \left(1 + \cos \pi \, \frac{V}{V_{\lambda/2}} \right)$$

(9.5)

where:

I = the light beam intensity at the detector,

C = a constant that reflects the efficiency of the detection including attenuation losses in the optical track,

V = modulating voltage:

$$V = Eh_{eff} T = Eh_{eff} \frac{C_A}{C_A + C_m}$$

(9.6)

If we assume here the typical magnitudes of applied power and reasonably measurable values of V and the modulator parameters, then we will have the sensitivity of the device, i.e., the minimal EMF intensity that could be measured with its use. The estimation, completed for the most commonly used modulator with LiNbO$_3$ crystal, gives a result somewhat below 100 V/m for wideband detection.

The estimation result is on the level of two orders of magnitude below those experimentally confirmed for similarly sized traditional probes. Although, using narrowband detection, it is possible to obtain sensitivity at a level of millivolts per meter, taking into account other advantages of the photonic probes, it is necessary to initiate further measures to increase its sensitivity. The most evident approach here, as mentioned above, is the development and the application of new crystals having lower values of the $V_{\lambda/2}$ voltage. This approach has been undertaken, and there are already known crystals that allow for better sensitivities as compared to lithium niobiate, although their availability is currently very limited. Below we will present two solutions that allow a sensitivity increase through the use of specific circuitry. Of course, they may be applied with an arbitrary type of crystal.

A. The Modulator with Multiple Access

The phase change δ observed at the interferometer's output is the result of the difference of the refractive indexes in the reference track of the interferometer and that in the modulated arm and is caused by the applied voltage that we will specify in the form:

$$\delta = \frac{2\pi}{\lambda_0} \left(n_m - n_0 \right) L$$

(9.7)

where:

    $L =$ the modulators length,

    $n_0 =$ the refractive index without modulation,

    $n_m =$ refractive index while a modulating voltage is applied.

As far as the influence of the refractive index variations on the phase difference reflects the evident role of the crystal (its sensitivity) applied in the modulator, the length of the modulator, which appears in Eq. (9.7), suggests the possible use of large-size modulators so as to obtain the probe's required sensitivity. As already shown in Chapters 4 and 5, the sizes are rigorously limited, primarily as a result of measured field averaging, particularly while the measurements are performed at distances to a source comparable to or less than the size of the probe. The perfect probe for our purposes here would be a zero-dimensional probe.

In the case of the magnetic field, a good approximation of the ideal is a probe with a Hall cell or with a magnetodiode, but their other weaknesses radically limit the use of these transducers as compared to loop antennas. Thus, although the increase of the modulator's size in the photonic probe allows a certain increase in its sensitivity, it also leads to limitations in its use, especially in close proximity to radiation sources. It also requires some limitation in the measuring frequency band since, apart from the resonant phenomena, the modulator works most effectively when measured field phase changes along the modulator are as small as possible.

An artificial increase in the modulator's length, without necessary changes in its geometrical dimensions, creates the possibility of multiple passage of the modulated light beam throughout the modulator. To illustrate the possibility, Fig. 9.9 shows three variants of this solution.

In Fig. 9.9a, we see the simplest solution of double passage of the light beam with its reflection at one end of the modulator and its withdrawal from the modulator through a separate (if necessary) fiber optic from the entry side. Because of the attenuation of the light beam in the electro-optical crystals, the interfering influence of the multiple reflected rays upon the resultant phase may be neglected. In Fig. 9.9b, a modulator with the multiple paths is shown in which the light beam is parallel to the optical axis of the crystal. In this way, a relatively good independence from the multipath propagation within the crystal is obtained as well as the most advantageous conditions of beam modulation. A "multiplied" version of the solution in Fig. 9.9a is shown in Fig. 9.9c. Two opposite sides of the modulator were equipped with reflecting surfaces (mirrors). The input and the output optical fibers are placed in the corners, free of the reflecting metallization, and aligned in such a way as to obtain the required number of beam passages throughout the modulator [8].

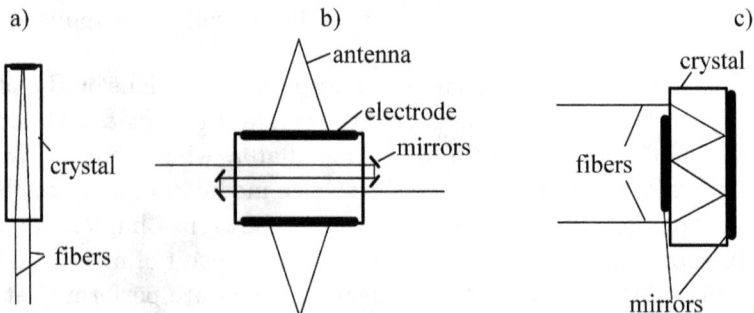

Figure 9.9 Multiple access modulators.

B. The Heterodyne Sensor

Two lasers, excited optically from a third one, create the heterodyne EMF probe shown in Fig. 9.10 [9]. This excitation by a third antenna is to ensure galvanic insulation of the probe against other parts of the measuring device and to feed the probe throughout the "transparent" line. One of the lasers is modulated in frequency by a voltage fed from the measuring antenna to an electro-optic crystal immersed inside the laser's resonator. The other laser plays the role of the heterodyne, and its modulator is used for the possible mean value of the differential frequency stabilization that allows a compensation for its instabilities. Frequencies of both lasers are mixed, and the differential signal, after detection, is fed to an indicating device. The heterodyne's differential signal is FM modulated and can be demodulated as a typical FM radio signal.

The concept shown here appears to be extremely attractive because of the possibility of achieving sensitivities exceeding those of traditional probes. The main problem here is stabilization of the devices operation. In order to improve thermal stability, both lasers are as identical as possible. They are placed close to one another under the same external conditions (temperature, pressure, humidity, vibration and acoustic noise, and light). The results of thermal creep, the laser-exciting power variations, and other factors affecting stability are compensated using a heterodyne modulator, which is also applied for setting the working point (differential frequency) of the system and to introduce feedback. The modulator is fed by a separate optical fiber. This does not address the fact that the measured frequency range in this solution may be more rigorously limited by difficulties with wideband FM detection than by the above-discussed phenomena.

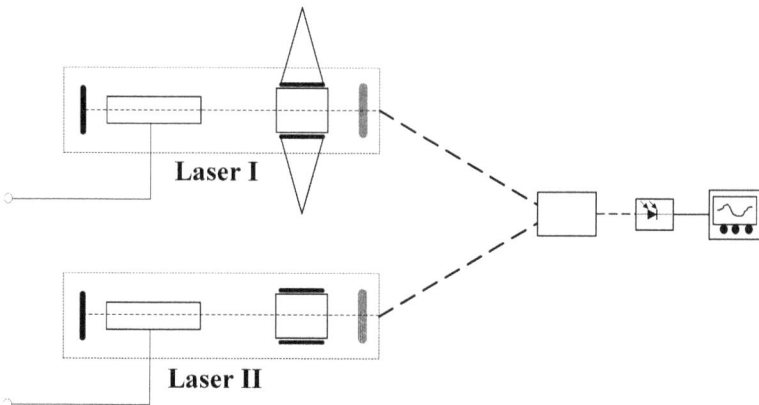

Figure 9.10 Block diagram of the photonic heterodyne EMF probe.

In Fig. 9.11, we see the measured spectrum of the FM modulated output of the investigated model of a laser heterodyne with $LiNbO_3$ lasers in the case of a 500 kHz modulating frequency and two voltages of modulating signals: 3 and 10 V.

Traditional approaches to FM modulated signal detection (frequency discrimination) may be used while the frequency deviation does not exceed, say, 1 GHz, which illustrates the maximum frequency at which

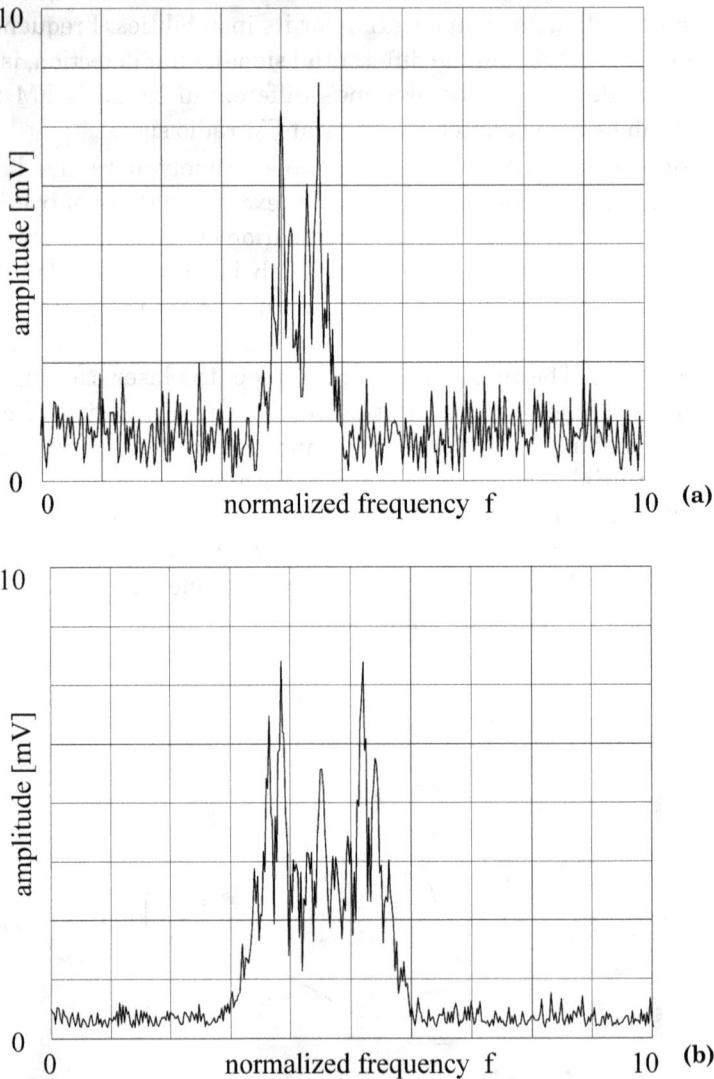

Figure 9.11 Output spectra of a heterodyne probe with $LiNbO_3$ modulator for two modulating voltages, (a) V = 3V and (b) V = 10 V.

the technique may be applied. In our case, we are interested in EMF measurements in as wide a frequency band as possible. That requires a new approach to frequency discrimination. Figure 9.12 shows a block diagram of an EMF FM probe in which an optic filter (OF) instead of a resonant circuit or a cavity is applied as a frequency discriminator.

The resonant frequency of the optical filter is selected so that the frequency of the FM modulated laser is approximately equal to that of half of one of the slopes of the filter. This allows a conversion of FM to AM in a similar manner as in the case of any other filter, with the exception of the much wider frequency band in which the discriminator works. Unfortunately, the wideband characteristic was achieved at the expense of sensitivity. But this should not come as a surprise, as a well known rule of basic electronics says that the product of amplification and bandwidth remains constant.

The design shown in Fig. 9.12 may be equipped with extra modulating electrodes, as in an FM modulated laser working as the filter for better linearity and stabilization of the probe's operation. This concept is mentioned in Section 9.5.

9.4 MAGNETIC FIELD PHOTONIC PROBE

As already mentioned, the construction of a photonic H-field probe is identical to that of an E-field probe, with the difference consisting of the use of a H-field antenna instead of an E-field one. In a "no-antenna" solution, a magneto-optic crystal should be used instead of an electro-optic one. Figure 9.13 shows a block diagram of an H-field meter in which a doubly loaded loop antenna is used. An exciting power from a laser source is split in the light divider and fed to two identical modulators located at two reverse sides of a loop antenna. After detection, the signals from both tracks are combined in similar manner as described

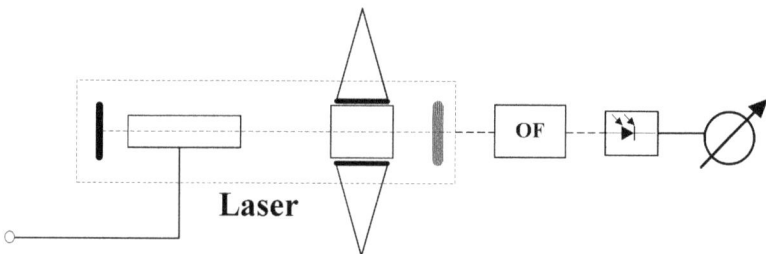

Figure 9.12 Block diagram of an EMF FM probe with optical frequency discriminator.

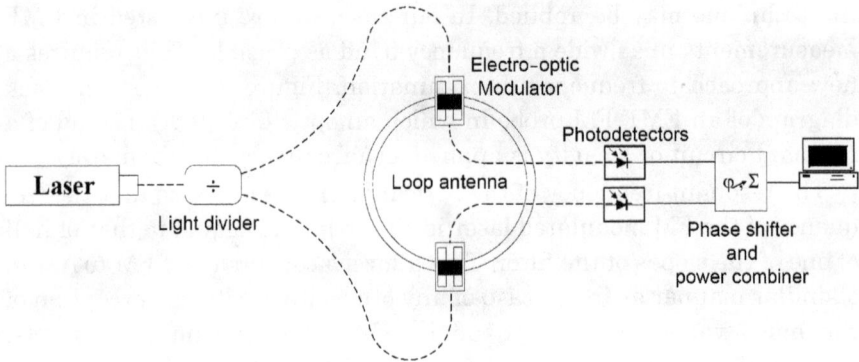

Figure 9.13 Block diagram of the H-field meter with doubly loaded loop antenna.

in Chapter 6. Here we are interested in a way to create a sensitivity increase in the H-field probe; however, much as before, the set allows E-field measurements as well.

9.5 DETECTOR LINEARITY

As can be seen from the shape of the dynamic characteristics of the interferometer's detector (and a similar one in the polarimetric system) as shown in Fig. 9.5b, the linear part of its characteristics is limited to only the range in which the function is represented by the first term of its expansion in the power series. Moreover, the output signal of the interferometer is a periodic function, and when the modulating voltage exceeds $\pm V_{\lambda/2}/2$, the output voltage decreases. Thus, the linear measuring range of the photonic probe (modulator) is relatively narrow, which causes problems with achieving required sensitivity on one hand, and problems with the linearity of the detector and the limitation of the maximal values of the modulating voltage (and the measured field) on the other. The sensitivity improvement methods were proposed above. In order to linearize the device and exclude unwanted consequences of the detector's periodic properties (i.e., to extend the range of measured values), an approach to compensation has been devised (Fig. 9.14) for the meter.

The modulator presented in Fig. 9.9b was designed and completed so as to make it possible to modulate all the transitions with one common voltage or separate excitation of all of them. One of the pathways may be used for negative feedback. Thus, it is fed by an output voltage from an amplifier excited by a detector. The voltage, which is simulta-

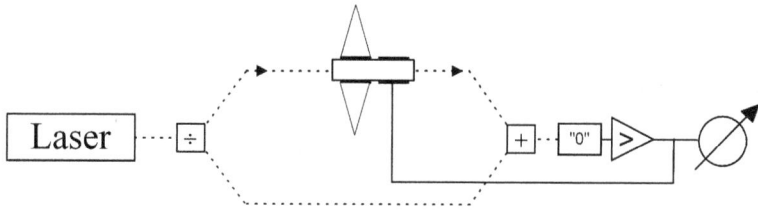

Figure 9.14   Compensational EMF measurement.

neously fed to an indicator, as a measure of the investigated field strength, is equal in magnitude to the modulating one, while its phase is opposite. In this way, compensation of the modulating signal is achieved. The compensational approach makes it possible to suppress the problems with modulator linearity and makes it possible to measure much larger field strength [10]. A disadvantage of the device is a certain limitation of its measuring frequency band resulting from the necessity of using fast amplifiers with high amplification and large dynamic range in the feedback (their dynamics limit those of the measured field). Aside from this, the concept may be helpful even in a limited frequency range when a wide dynamic range of the measured field is required.

An extension of the concept, based on the use of additional modulating electrodes in the reference track of the interferometer, permits use of interferometer "electric tuning" and feedback for stabilizing its working point [7].

9.6 SYNTHESIS OF THE SPHERICAL DIRECTIONAL PATTERN

Because of the described considerations, the output voltage of an optical detector precisely reflects the spectrum of the measured field (signal). If frequency components of different polarizations appear in the spectrum, and if we would like to take them all into account while interpretation of the measured field is prepared, the use of an omnidirectional probe will be indispensable. Let's consider several problems specific to the spherical pattern synthesis of the photonic probes.

Using methods similar to those presented in Chapter 7 and taking into account linearity of the photonic probe, it is possible to design an omnidirectional probe composed of three identical and linearly independent probes of a sinusoidal pattern (as has already been described in Chapter 7) and then summing squares of their output voltages. This method is evident and efficient; however, it requires use of a relatively complex measuring system. An essential issue arises here: the result of summation of the squared spatial components of spherically polarized signals is a constant value. As described in Chapter 7, without regard to how the measured signal was interpreted, this was a DC component. The matter was in the DC component to the extent to which it was the subject of the measurement; i.e., the component was proportional to the intensity of the measured E- or H-field component. Meanwhile, in the considered case, the (optical) detector is loaded with a spectrum analyzer, a selective HF millivoltmeter, an oscilloscope, or another device for the alternating voltages' measurement.

The constant value of the output voltage, provided in considerations of the spheroidal polarized fields measurement, is the result of filtering of the output signal as well as the large time constants of the detectors. In the measured field, there exists one and only one linearly polarized vector of the E-field and one of the H-field at any instant of time. We should notice that in the optical detector there are no filtering (averaging) elements and, as a result, it is "fast," and the output voltage of the detector is an AC voltage, proportional to the instantaneous field strength that represents the amplitude, phase, and frequency of the field.

The construction of an omnidirectional probe using traditional technology (an antenna loaded with a diode) was relatively simple, as it was assumed (with an acceptable accuracy and within limited dynamic range) that the characteristics of the diode are square-law, and the summation of the output voltages of three mutually perpendicular probes was enough to obtain the required omnidirectional properties. In the case of photonic sensors, a similar approach is very possible. However, it is necessary to use probes (modulators) whose working point was selected for $\varphi_0 = 0$ (Fig. 9.5a) to achieve this. Their output voltage is proportional to the square of the modulating voltage. Thus, by way of parallel connection of three identical probes (Fig. 9.15), positioned in three mutually perpendicular directions and sensitive to three different spatial components of the measured field, it is possible to synthesize, in a relatively simple manner, the omnidirectional pattern. A similar result may be obtained when the probes are in serial connection; however, in this case, their working point must be shifted by π in relation to the previous one. Another version of the solution shown in Fig. 9.15 may be realized by changing the order of the summation and the detection. Assuming the detector to be a linear device, we will have exactly the same result, and its additional advantage will be the possibility of independent measurement of the spatial components of the investigated field. Unfortunately, as may be seen from the curves shown in

a)

b)

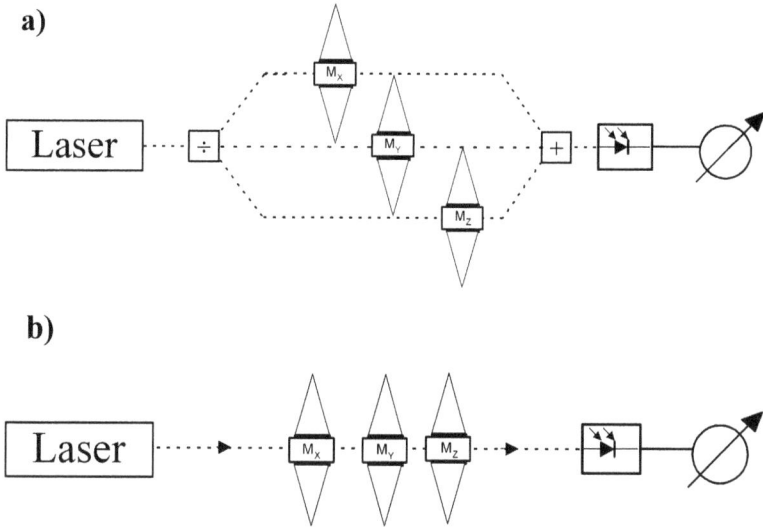

Figure 9.15   The omnidirectional photonic probe with (a) modulators in parallel and (b) serial connection.

Figs. 9.5a and 9.5b, simplification of the omnidirectional probe construction was obtained at the expense of a remarkable reduction in its sensitivity. But, even here, the attractiveness of the simpler construction sometimes may be crucial [11].

9.7   THE "FUTURE METER"

Current-day meters used for workplace safety and general public protection meet the demands formulated for them, especially in applications specific to surveying and monitoring services. Apart from the disadvantages of the meters, presented in previous chapters, their most important imperfection is the possibility of being misled by identical meter indications when fields of different parameters (frequency spectrum, modulation type, polarization) are measured and a lack of precisely defined magnitudes characterizing the field, which should be measured. These unclear measurement conditions may lead to investigation results that are not equivalent in interpretation, not only in laboratory conditions, but in epidemiological studies, for example. This makes the comparison of exposure data of selected populations from different measuring teams, using a variety of measured equipment, impossible. The problem becomes particularly critical when steps are taken toward creating a complex evaluation and comparison of the professional or/and nonprofessional exposure of selected groups of people and those of everybody involved.

It seems that the most sophisticated measuring demands are fulfilled by a meter equipped with wideband active antennas connected to a spectrum analyzer and then to a computer. Currently applied active antennas require their connection with a measuring system via a coaxial cable, which is not permissible when measurements are performed in the near field. Using the discussed photonic approach here, it is possible to construct the set using the block diagram shown in Fig. 9.16.

This shows three mutually perpendicular magnetic and electric field probes connected through a sampling and multiplexing system to a spectrum analyzer controlled by a computer. Every spatial component of any frequency component is measured by the automatically controlled spectrum analyzer, which allows a complex analysis of the field. Contrary to the meters available on the market whose frequency response is matched to specific national or other standards the use of data processing allows an arbitrary interpretation of the measurement results. Flexible selection of the optional software becomes decisive here instead of a meter frequency response set by a manufacturer, with no possibility of changing it anytime during the meter's use. The software flexibility leads to the possibility of using antennas (probes) of an

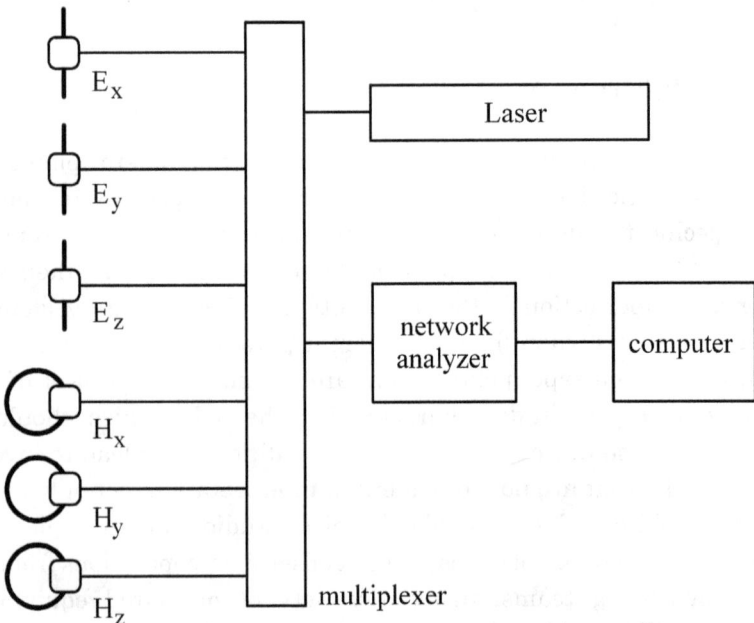

Figure 9.16 Block diagram of the "future meter."

arbitrary but known frequency response. As a result, management of the measured results is simple here, analysis in an arbitrarily selected aspect is possible, and standardization of the data interpretation methodology does not create problems in archiving the selected measurement results. The presented option, apart from the possibilities described, allows observations of temporal variations of the investigated field and may play the role of a dosimeter, i.e., a device counting the product of time and the measured value (dose) [12].

The construction of such a device is already possible, and the first steps toward its completion have already been taken [12, 13]. However, in the future, due to a projected increase in the development and accessibility of photonic technologies and the usual decrease in prices, a device with this structure may become a fundamental tool for monitoring services as well.

It may be supposed that these meters will be available in different versions, some even more advanced. For instance, they might be equipped with more probes connected to one measuring device for spatial field distribution investigations or for gradient studies. Of course, versions destined for general use should be simplified in the sense of their construction, measuring possibilities, and maintenance and, as a result, will be less expensive.

As already mentioned, measuring devices used today, although loaded with many disadvantages and inconveniences, may satisfy nearly all contemporary measurement requirements. A significant part of these negative features will remain in the most "intelligent" solutions as well. We know that the practical ability to use relatively simple meters presently creates many problems, especially for (but not limited to) non-technicians. We therefore should expect and accept that the use of devices offering a much wider range of measuring capabilities, which means more sophisticated ones, will force users to sharpen their measurement skills.

References

[1] E. Grudzinski, R. Kinda, J. Poreba, Z. Siwek. Remote EMF strength measurements above earth surface (in Polish). *Proc. Natl. Telecomm. Symp.* KST-89, pp. 330–335.

[2] E. Grudzinski, H. Trzaska. General public protection against electromagnetic radiation. *Proc. Intl. EMC Symp.* Nagoya 1989, pp. 742–746.

[3] M. L. Van Blaricum. Photonic systems for antenna applications. *IEEE AP Magazine,* vol. 36, No. 5, Oct. 1994, pp. 30–38.

[4] M. Kanda. Optically sensed EM-field probes for pulsed fields. *Proc. of the IEEE,* vol. 80, No. 1, Jan. 1992, pp. 209–214.

[5] T. Babij, H. Trzaska. Properties of wideband magnetic field probes. *Proc. 1976 IEEE Intl. EMC Symp.,* Wash. DC, pp. 375–380.

[6] E. R. Mustiel, W. N. Parygin. *The light modulation methods (in Russian).* Svyazizdat, Moscow, 1973.

[7] P. Bienkowski, H. Trzaska. Frequency limitations in photonic EMF probes. *Proc. 1997 Intl. EMC Symp.* Zurich, pp. 603–606.

[8] H. Trzaska Photonic electromagnetic field probes. *Proc. EMC 1996 Roma Intl. Symp.,* pp. 221–226.

[9] P. Bienkowski, H. Trzaska. The new approach to the photonic EMF measurements. *Proc. Intl. EMC Symp.* Wroclaw 1996, pp. 347–350.

[10] P. Bienkowski, H. Trzaska. New EMF photonic sensors. USNC/ URSI Meeting, Boulder, CO, 1996.

[11] S. Diba, H. Trzaska. Isotropic receive pattern of an optical EMF probe based upon a Mach-Zehnder interferometer. *IEEE Trans.* vol. EMC-39, No. 1, 1997, pp. 61–63.

[12] P. Bienkowski, H. Trzaska. Photonic EMF measurements. XXV-th General Assembly of the URSI, Lille, France 1996, Abstract p. 581.

[13] K. Tajima, R. Kobayashi, N. Kuwabara. An improved broad-band isotropic E-field sensor using Mach-Zehnder interferometers. *Proc. 1999 Intl. EMC Symp.* Zurich, pp. 555–559.

Chapter 10

Final Comments

This book discusses EMF measurement possibilities in the near field, especially for human safety and environmental protection purposes. Field measurement technical problems in the near field are similar or identical to those in the more widely understood area of electromagnetic compatibility. Some differences between the far and the near field were emphasized, especially those essential to measurements performed in the near field. EMF measurement methods in the near field were presented, and details were provided about factors limiting accuracy of the electric- and magnetic-field and power density measurement in the near field using small electric and small geometric antennas. The influence of external factors, which are important to estimation accuracy, was outlined, and new measuring possibilities, based mainly on the development of photonic techniques, were briefly presented.

It seems that the considerations presented herein allow us to formulate three summarizing questions:

1. The presented estimations are analytical in nature. Is it possible to perform a synthesis that would allow the final, synthetic conclusion with regard to measurement accuracy? The estimation presented is only an example.

2. What is the relationship between the presented material and today's legal regulations, and how should future regulations be directed?

3. What are the perspectives of development in the field?

We will try to briefly present the authors' views relating to the above-formulated questions and doubts surrounding these questions. These subjective opinions have resulted from the authors' knowledge and experience as well as the nature of the presented work. (This work was prepared as a sort of introductory selection, summary, and closed-form assembly of present knowledge in the field and provided as a background for wider discussion and indispensable complementation that may take place within the framework of one or more competent body concerned in the field. We may add here that these bodies shoot up like mushrooms, and a question may sometimes arise as to whether there are too many of them.) Taking into account the possibility of preparation, on the grounds of the presented material, a wider elaboration on the present status of the near-field EMF metrology, any comments and suggestions from readers on the matter are invited by the authors, and they are appreciated very much in advance.

An estimation of the entire subject of measurement accuracy, taking into consideration the specificity of measurements for human safety and environmental protection purposes, seems to be both impossible and unnecessary. The impossibility may be substantiated by the undefined *a priori* object of the measurements. Let's remember that, in many cases, the primary aim is to reveal and identify the source of radiation and then just to measure field generated by it. It is only when the location of the primary source and other objects having an effect on the field generated by the source are identified, thus *a posteriori,* that precise considerations of the accuracy of the performed measurement may be completed, and it may be done only if the geometry of propagation is well known and exclusively for defined point in space where the probe is placed. On other hand, the necessity of accurate determination of the measured quantity depends on the purposes of the measurement. In our case, we are particularly interested in finding maximal values in the sense of time and space. However, these values may be characterized by rapid variation that, in many cases, may well exceed the inaccuracy of the applied measuring devices. For example, in the process of dielectric welding, the field intensity near the device changes two to five times during a single welding cycle whose duration does not exceed several seconds. It is impossible to predict what, under these conditions, will be read by a person performing the measurement. Probably, the measured magnitude will be read near the end of the cycle, when the field becomes almost constant, but it is unknown whether it was maximal, minimal, or something in between. All of it is done with a meter with an estimated measuring error, say, 10 percent. The procedure may be described with the use of an old expression, "breaking a butterfly upon a wheel."[*] This does not mean, of course, that the inaccuracy of the measurement may be arbitrary.

---

[*] From Alexander Pope's "Epistle to Dr. Arbuthnot," January 1735. The "wheel" refers to a "breaking wheel," a device used for torture in the Middle Ages.

The presented considerations of accuracy-limiting factors show that the most essential factors are the distance between a source and a measuring probe, the sizes of the measuring probes, and the thermal effects. The presence of any objects that cause deformation of the measured field within the measurement area must be included in considerations about accuracy risks, and understood possibly requiring spatial reconfiguration for technological, operational, or other reasons. A set of separate problems is created by pulsed-field measurement. In order to avoid the problems mentioned (and others), an accurate assessment of the parameters of the equipment used, its advantages and weaknesses, as well as its correct use, is required from the services performing the measurements. It may be said, with some approximation, that quite good measurements may be performed with the use of a relatively poor device if its parameters, as well as the measurement conditions, are known and understood. A separate and final factor in the accuracy of a measuring device is the accuracy of its calibration. This usually does not exceed 0.5 to 1 dB.

The legal issues in this field play a very important role, because they are valid. However, most national standards, and in particular international regulations and recommendations, are "rigid." Amending, revising, and adopting standards to reflect the newest trends and knowledge is, at best, time consuming. An outstanding example is the ANSI standard, which is periodically reviewed and revised. It is the child of international regulations, with many different organizations involved (e.g., WHO, ITU, IEC, URSI, EC). Perhaps because of this, revisions are difficult, as noted by Dr. Kunsch [4]. It is the authors' hope that it at least should be possible to establish one competent international body to prepare a reasonable proposal. The community in this field is relatively small (as compared to many other fields), and it is grouped more or less officially around the Bioelectromagnetics Society (BEMS) and the European Bioelectromagnetics Association (EBEA) [5]. Their members dominate, in different combinations, in the regulatory bodies mentioned.

The authors, as technicians, do not reserve for themselves any right to decide what threshold limits should be accepted in future standards. In their opinion, presented many times before, the decision must be reserved to medical personnel, and the role of physicists and engineers should be an auxiliary one. Of course, participation of the latter in any working group and research team is indispensable as, similar to technicians in biology, the biologists and medical people have certain difficulties with technical and physical problems. The standards worked out this way may be a bit less accurate, but they will be much more humanistic. On the contrary, the limits proposed by technicians, in the form of model studies (even "millimeter resolution" ones) may be accepted as introductory proposals but nothing more. Their approach has nothing in common with phenomena *in vivo,* where proposed limits are given with an accuracy remarkably exceeding the repeatability of biomedical experiments and the accuracy of available field meters. This point is well illustrated by the limits shown in Tables 1.4 and 1.6, where they are given with an accuracy to three or more significant figures! This only confirms the mechanistic, unacceptable character of these proposals. Disregarding differences between Western and Eastern standards, one of the drawbacks of existing legal acts, proposals, recommendations, and so on is the absence of any information of a metrological character, and these data are excessively detailed [2,3]. The latter approach, especially in an international recommendation where, due to long and complex bureaucratic coordination procedures, any change or modification may be more difficult than its original adoption, limits or excludes any further improvements of the existing methods or implementation of new methods that could assure much better results (in technical, operational and interpretational sense) than methods (instruments) presently in existence.

The same may be said with regard to calibration methods. In the opinion of the authors, any protection standard should include a short statement addressed to the metrology. It should contain only the most important parameters, and it may be formulated, for instance, in the form:

> *The subject of the measurement is the RMS value of the measured quantity, measured in the extreme conditions within defined frequency ranges. Required accuracy of the measurement: better than 1 dB when measurements are performed at a distance 5 cm from the nearest radiating source (primary or secondary one) or other material media.*

The statement contains all the information necessary for correct measurements. Because of the possibility and necessity of measured data comparison and use for statistical, epidemiological, and other studies, the acceptance of a common measurement methodology is necessary. The problem may be solved by attaching a statement similar to the above to a protection standard. In this manner, a universal, short, and clear document may provide all information necessary for surveying and monitoring services.

Development perspectives in this field are almost unlimited. It is worth mentioning, as an example, the need to measure fields generated by living organisms, but these measurements require sensitivities well in excess of what is now achievable. These possibilities were partially described in Chapter 9. However, the conclusion formulated there is still valid—that the newest solutions are designed primarily for research laboratories and specific institutions. Whereas basic equipment for surveying and monitoring services probably will not change significantly from that presently employed, over the next dozen or so years. It may be expected that the market will bring forth meters that allow simplified measurements. However, the measurement techniques and methodology are excessively complicated even now, and it is often not understood, even by many people with quite good metrological experience. According to the authors' estimation, more than 50 percent of biomedical experiments are performed under conditions that are unacceptable from a technical point of view. It is not surprising that, in field measurements, experienced metrologists use EMF meters equipped with loop antennas for near-field E-field measurements because "*they are calibrated in the E-field units!*"

Another experienced metrologist tried to suggest that he had measured field levels generated by a satellite TV receiver that well exceeded those permitted by standards; it was actually the secondary field generated in an ungrounded feeder of a roof converter by a broadcast FM station located in the neighborhood. Similar tales may be told. They confirm that far-field experience is not enough in near-field measurements and that additional training is necessary. The statement can be applied both to meter manufacturers and meter users. Because of the variety of meters available on the market, choosing an appropriate meter may sometimes be problematic, especially for non-technicians. For instance, not every attractively priced meter is worth our attention. To illustrate the wide range of proposed designs and, on one hand, their metrological possibilities and, on the other, difficulties with the decision "what to choose" Table 10.1 presents selected probes and meters designed and available in Poland for our discussed purposes. The table does not imply that these meters are the best meters in the world; it only shows what is available to Polish services, the needs of their users, and characteristics of the manufacturers. By the way, it also illustrates the necessity of matching of the measuring devices to the country's standards.

Based on these discussions, we can try to formulate the basic parameters that characterize a meter's applicability in near-field measurements and what should be taken into account when a meter is evaluated or when a meter is to be chosen to meet specific requirements:

1. Measured quantity (Chapter 3)
2. Probe sizes (Sections 4.1, 4.3, 5.1, and 5.4)
3. Frequency response (Sections 4.2, 5.2, and 5.5)
4. Directional properties of the probe (Sections 4.4, 5.3, and 7)
5. Measured value: peak, mean, or RMS (Section 8.2)
6. Dynamic range (Section 8.2)
7. Required measurement accuracy
8. Thermal stability (Section 8.1)
9. Possible aging effects (mentioned in Chapter 8)
10. Immunity to external EMF (Section 8.4)
11. Measured field deformations (Sections 8.4 and 8.5)
12. Complexity of design, operation, and service
13. Endurance, reliability, price, and cost of operation
14. Conformability with national or international standards or specific requirements of the planned measurements
15. Manufacturers credibility, solidity of a dealer, and accessibility for maintenance
16. Calibration accuracy and periodic recalibration capability

The authors would like to express their gratitude to the editors for their hard work in adapting the authors' "Penglish" (Polish-English) version of the manuscript into more comprehensible language.

References

[1] ICNIRP Guidelines. Guidelines on limits of exposure to static magnetic fields. *Health Physics,* No. 1/1994, pp. 100–106.
[2] Measuring equipment for electromagnetic quantities. Prepared by IEC TC 85 WG11.
[3] Radio transmitting equipment. Measurement of exposure to radio frequency electro-magnetic field–field strength in the frequency range 100 kHz to 1 GHz. IEC SC12C.
[4] B. Kunsch. The new European Pre-Standard ENV 50166—Human exposure to electromagnetic field. COST 244 Working Group Meeting, Athens 1995, pp. 48–58.
[5] ICNIRP Guidelines. Guidelines on limits of exposure to static magnetic fields. *Health Physics,* No. 1/1994, pp. 100–106.

Table 10.1 Selected Meters and Probes Used in Poland

Meter	Probe	Field	Pattern	Frequency Range	Measuring Range
MEH Technical University of Wroclaw, Poland	AE 4	E	s	10–1000 Hz	0.1–30 kV/m
	AE 3	E	s	1–1000 kHz	5–1000 V/m
	3AE 1	E	o	0.1–300 MHz	2–1000 V/m
	3AS 1	S	o	0.3–3 GHz	0.1–150 W/m^2
	3AS 3	S	s	0.3–40 GHz	0.1–150 W/m^2
	AH 4	H	s	40–1000 Hz	1–500 A/m
	AH 3	H	s	1–100 kHz	1–250 A/m
	3AH 1	H	o	0.1–10 MHz	1–250 A/m
	AH 27	H	s	10–100 MHz	0.01–10 A/m
HI-2200 Holaday – ETS Lindgren	E100	E	o	0.1–5000 MHz	0.3–800 V/m
	H200	H	o	5–300 MHz	0.03–10 A/m
	II210	H	o	0.3–30 MHz	0.3–30 A/m
NBM-500 STS Narda	EF-0391	E	o	0.1–3000 MHz	0.3–444 V/m
	EF-6091	E	o	0.1–60 GHz	0.3–30 A/m
	HF-3061	H	o	0.3–30 MHz	0.02–16 A/m
	HF-0191	H	o	27–1000 MHz	0.02–16 A/m
PMM 8053 PMM Italy	EP-408	E	o	0.001–40 GHz	1–800 V/m
	EP-300	E	o	0.1–3000 MHz	0.1–300 V/m
	HP-032	H	o	0.1–30 MHz	0.01–20 A/m
	HP-102	H	o	30–1000 MHz	0.01–20 A/m
ESM-100 Maschek	–	E	o	5 Hz–400 kHz	1 V/m–100 kV/m
	–	H	o	50–1000 Hz	10 nT–10 mT

Indications:
E = electric field   p = peak value probe
H = magnetic field   s = sinusoidal
S = power density   o = spherical

Index

A
accuracy
  dependence on input impedance 71
  dependence on source 69
  estimation 85, 114
  limitations 3, 41, 54, 89, 209
  magnitude of error 111
  measurement 7
  of EMF standards 3
  photonic probes 186
  power density measurement 119
  pulsed EMF measurement 163
  vs. distance from source 100
American Conference of Governmental Industrial Hygienists 5
ANSI/IEEE C95.1–1992 5
antenna effect 89, 90, 122
antennas
  asymmetrical 85
  electric 32
  input impedance changes 71, 102
  Kathrain GSM900 125
  large power 185
  multiple-loaded loop 91
  quadrant 138
  radio station 18
  rotating radar 163
  size 54, 75, 89, 126
  symmetrical dipole 51
  types 81

B
Bessel function 61, 96
Bioelectromagnetics Society 209
Biot-Savart law 16, 46

C
circular polarization 142
Coulomb's law 16

D
direct current summation 142
directional pattern 74, 95
Doppler effect 19, 133

E
EMF 51, 75, 100, 138, 144
  exposure limits 4
  modulation 31
  polarization 30
  spectrum 28
  units 4
  variations 24
energy absorption 48
European Bioelectromagnetics Association 209
European regulations 4

F
Faraday's law 100
field
  averaging 52
  delimitation 15
field strength measurement 54
filters
  choice of corner frequencies 85
  low-pass 64
  multiple 59
  multi-segment 61
  notch 63
  RC band-pass 128
  RC low-pass 55
  sum of capacitances 62
  sum of reactances 72
  use of additional RC 95
folded dipole 94
Fraunhofer region 3, 16
frequency response
  magnetic field probes 92
  meters 29